T0224867

Lecture Notes in Mathematics

Edited by A. Dold, F. Takens and B. Teissier

Editorial Policy
for the publication of monographs

1. Lecture Notes aim to report new developments in all areas of mathematics – quickly, informally and at a high level. Monograph manuscripts should be reasonably self-contained and rounded off. Thus they may, and often will, present not only results of the author but also related work by other people. They may be based on specialized lecture courses. Furthermore, the manuscripts should provide sufficient motivation, examples and applications. This clearly distinguishes Lecture Notes from journal articles or technical reports which normally are very concise. Articles intended for a journal but too long to be accepted by most journals, usually do not have this "lecture notes" character. For similar reasons it is unusual for doctoral theses to be accepted for the Lecture Notes series.

2. Manuscripts should be submitted (preferably in duplicate) either to one of the series editors or to Springer-Verlag, Heidelberg. In general, manuscripts will be sent out to 2 external referees for evaluation. If a decision cannot yet be reached on the basis of the first 2 reports, further referees may be contacted: the author will be informed of this. A final decision to publish can be made only on the basis of the complete manuscript, however a refereeing process leading to a preliminary decision can be based on a pre-final or incomplete manuscript. The strict minimum amount of material that will be considered should include a detailed outline describing the planned contents of each chapter, a bibliography and several sample chapters.
Authors should be aware that incomplete or insufficiently close to final manuscripts almost always result in longer refereeing times and nevertheless unclear referees' recommendations, making further refereeing of a final draft necessary.
Authors should also be aware that parallel submission of their manuscript to another publisher while under consideration for LNM will in general lead to immediate rejection.

3. Manuscripts should in general be submitted in English.
Final manuscripts should contain at least 100 pages of mathematical text and should include
– a table of contents;
– an informative introduction, with adequate motivation and perhaps some historical remarks: it should be accessible to a reader not intimately familiar with the topic treated;
– a subject index: as a rule this is genuinely helpful for the reader.

Lecture Notes in Mathematics 1736

Editors:
A. Dold, Heidelberg
F. Takens, Groningen
B. Teissier, Paris

Springer
Berlin
Heidelberg
New York
Barcelona
Hong Kong
London
Milan
Paris
Singapore
Tokyo

Bengt Ove Turesson

Nonlinear Potential Theory and Weighted Sobolev Spaces

 Springer

Author

Bengt Ove Turesson
Matematiska istitutionen
Linköpings Universitet
SE-58183 Linköping, Sweden

E-mail: betur@mai.liu.se

Cataloging-in-Publication Data applied for

Die Deutsche Bibliothek - CIP-Einheitsaufnahme

Turesson, Bengt Ove:
Nonlinear potential theory and weighted Sobolev spaces / Bengt Ove
Turesson. - Berlin ; Heidelberg ; New York ; Barcelona ; Hong Kong ;
London ; Milan ; Paris ; Singapore ; Tokyo : Springer, 2000
  (Lecture notes in mathematics ; 1736)
  ISBN 3-540-67588-4

Mathematics Subject Classification (2000): 31C45,46E35

ISSN 0075-8434
5ISBN 3-540-67588-4 Springer-Verlag Berlin Heidelberg New York

Springer-Verlag is a company in the BertelsmannSpringer publishing group.
© Springer-Verlag Berlin Heidelberg 2000

Typesetting: Camera-ready $T_E X$ output by the author
Printed on acid-free paper   SPIN: 10725034   41/3143/du   543210

Introduction

Let w be a weight on $\mathbf{R}^N$, i.e., a locally integrable function on $\mathbf{R}^N$ such that $w(x) > 0$ for a.e. $x \in \mathbf{R}^N$. Let $\Omega \subset \mathbf{R}^N$ be open, $1 \leq p < \infty$, and m a nonnegative integer. The weighted Sobolev space $W_w^{m,p}(\Omega)$ consists of all functions u with weak derivatives $D^\alpha u$, $|\alpha| \leq m$, satisfying

$$\|u\|_{W_w^{m,p}(\Omega)} = \left(\sum_{|\alpha| \leq m} \int_\Omega |D^\alpha u|^p w\, dx \right)^{1/p} < \infty.$$

In the case $w = 1$, this space is denoted $W^{m,p}(\Omega)$. Sobolev spaces without weights occur as spaces of solutions for elliptic and parabolic partial differential equations. Typically, $2m$ is the order of the equation and the case $p = 2$ corresponds to linear equations. Details can be found in almost any book on partial differential equations. For degenerate partial differential equations, i.e., equations with various types of singularities in the coefficients, it is natural to look for solutions in weighted Sobolev spaces; see, e.g., Fabes, Kenig, and Serapioni [36], Fabes, Jerison, and Kenig [35], Fabes, Kenig, and Jerison [37], and Heinonen, Kilpeläinen, and Martio [59].

A class of weights, which is particularly well understood, is the class of A_p weights that was introduced by B. Muckenhoupt in the early 1970's. This class consists of precisely those weights w for which the classical singular integral and maximal operators are bounded from the weighted L^p space $L_w^p(\mathbf{R}^N)$ to $L_w^p(\mathbf{R}^N)$, if $1 < p < \infty$, and from $L_w^1(\mathbf{R}^N)$ to the weak weighted L^1 space wk-$L_w^1(\mathbf{R}^N)$, if $p = 1$. Let us mention one example of these kinds of results. The Hardy–Littlewood maximal function, Mf, of a function f on $\mathbf{R}^N$ is defined by

$$Mf(x) = \sup_{r>0} \fint_{B_r(x)} |f(y)|\, dy.$$

Here, $B_r(x)$ is the ball with radius r and center in x and

$$\fint_{B_r(x)} |f(y)|\, dy = \frac{1}{|B_r(x)|} \int_{B_r(x)} |f(y)|\, dy,$$

where $|B_r(x)|$ is the N-dimensional measure of $B_r(x)$. A celebrated theorem by Muckenhoupt [85] from 1972 states that the Hardy–Littlewood maximal operator M, which takes f to Mf, is bounded on $L_w^p(\mathbf{R}^N)$, where $1 < p < \infty$, if and only if $w \in A_p$, which means that there exists a constant C such that

$$\left(\fint_B w \, dx \right) \left(\fint_B w^{-1/(p-1)} \, dx \right)^{p-1} \leq C$$

for every ball $B \subset \mathbf{R}^N$. For $p = 1$, the corresponding condition, which defines the class of A_1 weights, is that

$$\left(\fint_B w \, dx \right) \operatorname*{ess\,sup}_{x \in B} \frac{1}{w(x)} \leq C$$

for every ball $B \subset \mathbf{R}^N$. Working with A_p weights, it is thus possible to make use of well-known techniques from harmonic analysis. Another reason for studying A_p weights is the fact that powers of distances to submanifolds of $\mathbf{R}^N$ often belong to A_p. These kinds of weights are common in applications.

It is well-known that classical potential theory is connected to linear partial differential equations and the Sobolev space $W^{1,2}$. The most striking manifestation of this connection is the Dirichlet principle, which states that the solution to Dirichlet's problem for the Laplace equation in a domain Ω:

$$\begin{cases} \Delta u = 0 & \text{in } \Omega, \\ u = f & \text{on } \partial\Omega, \end{cases}$$

where the boundary function f is assumed to belong to $W^{1,2}(\Omega)$, can be obtained by minimizing the energy integral

$$\int_\Omega |\nabla u|^2 \, dx$$

over all functions $u \in W^{1,2}(\Omega)$ for which $u - f \in \overset{\circ}{W}{}^{1,2}(\Omega)$, the last space being the closure of $C_0^\infty(\Omega)$ in the norm of $W^{1,2}(\Omega)$. During the last forty years, a corresponding nonlinear potential theory, connected to nonlinear partial differential equations and the space $W^{m,p}$, has been developed. The theory originated in the work by V. G. Maz'ya and J. Serrin. Excellent accounts of this theory and its history are the monographs Adams–Hedberg [7], Maz'ya [76], and Ziemer [108]. A lot of the corresponding weighted theory can be found in Heinonen, Kilpeläinen, and Martio [59].

Below, we will mention a few of the concepts and ideas from potential theory that will be important to us. One central object of study in this theory is the concept of capacities. These set functions play a role for functions in $W_w^{m,p}$ which is similar to that of measures for functions in L_w^p, for instance, as a tool for measuring exceptional sets. One should think of a capacity as a finer tool than a measure in the sense that a set may have positive capacity, although it has measure zero. We will investigate several types of capacities, each with its own set of advantages: for $1 < p < \infty$, Bessel capacities, Riesz capacities, and variational capacities, and for $p = 1$, Hausdorff capacities and variational capacities. Of these, the variational capacity $c_{m,p}^w$ is directly associated with the norm in $W_w^{m,p}$. The capacity $c_{m,p}^w(K)$ of a compact set K is defined by

$$c_{m,p}^w(K) = \inf\{ \|\varphi\|_{W_w^{m,p}(\mathbf{R}^N)}^p \,;\, \varphi \in C_0^\infty(\mathbf{R}^N) \text{ and } \varphi \geq 1 \text{ on } K\}.$$

The capacity is extended to general sets as an outer measure. When $w = 1$, we shall denote this capacity by $c_{m,p}$. Another concept from potential theory, which we will study at some length, is the concept of thinness. This is connected with the problem of measuring the density of a set close to a point. We recall from classical potential theory that a point x on the boundary of a domain Ω is said to be regular for Dirichlet's problem for the Laplace equation if, for every continuous boundary function f, the solution u is continuous at x. Moreover, x is regular if and only if Ω^c is thick, i.e., not thin, at x. A quantitative definition of thickness is that the following integral diverges:

$$\int_0^1 \frac{c_{1,2}(\Omega^c \cap B_r(x))}{r^{N-2}} \frac{dr}{r} = \infty.$$

This is Wiener's criterion for regularity of a boundary point, which was proved by N. Wiener [107] in 1924. Similar conditions for the regularity of a boundary point for nonlinear equations were obtained by V. G. Maz'ya [72] in 1970. In particular, he showed that x is regular for the Dirichlet problem for the p-Laplace equation:

$$\Delta_p u = \text{div}(|\nabla u|^{p-2}\nabla u) = 0$$

whenever

$$\int_0^1 \left(\frac{c_{1,p}(\Omega^c \cap B_r(x))}{r^{N-p}} \right)^{p'-1} \frac{dr}{r} = \infty,$$

where $p' = p/(p-1)$ is the exponent conjugate to p. That this condition is also necessary was established by T. Kilpeläinen and J. Malý [68] in 1994. Let us finally mention the Kellogg property, which states that the capacity of the set of points in a given set, where the set is thin, is zero. Thus, except for possibly a set of capacity zero, every point on the boundary of a domain is regular. This result was proved by O. D. Kellogg [65], [66] in 1928 in the planar case, and by G. C. Evans [33] in 1935 for higher dimensions. Corresponding results for nonlinear capacities have been obtained by L. I. Hedberg [54] and Hedberg and Th. H. Wolff [58].

The purpose of these notes is to show how a number of results from the theory of non-weighted Sobolev spaces and from non-weighted potential theory can be extended to the weighted context. Almost all results are established for A_p weights; a few results are shown to hold for more general classes of weights such as doubling weights. The book is based on earlier work by the author in [103]. Unless otherwise stated, the results are due to the author (to the best of his knowledge).

As an application of the theory developed, we shall prove two simple theorems on "spectral synthesis" for weighted Sobolev spaces—one for the case $1 < p < \infty$ and one for the case $p = 1$. Our results cover the case when Ω is a bounded Lipschitz domain. The following spectral synthesis theorem is due to L. I. Hedberg [57], [58].

Theorem. *Let $u \in W^{m,p}(\mathbf{R}^N)$, where $1 < p < \infty$, and let $\Omega \subset \mathbf{R}^N$ be open. Suppose that $D^\alpha u = 0$ on Ω^c for $|\alpha| \le m - 1$.[†] Then u belongs to the closure of $C_0^\infty(\Omega)$ in $W^{m,p}(\mathbf{R}^N)$.*

Among the consequences of this theorem are uniqueness theorems for the Dirichlet problem for various elliptic differential operators; see Adams–Hedberg [7].

An early version of this theorem was proved in 1937 by S. L. Sobolev [96], who established the theorem in the case $p = 2$ and under the assumption that Ω is bounded by smooth manifolds. This result was used by Sobolev to prove a uniqueness theorem for the polyharmonic equation. With $m = 1$ and $p = 2$, the theorem was proved by A. Beurling [15] in 1947 and J. Deny [31] in 1950, and with $m = 1$ and $1 < p < \infty$ by V. P. Havin [52] in 1968 and T. Bagby [11] in 1972. For $m \ge 2$, partial results were obtained by J. C. Polking [93] in 1972, Hedberg [55] in 1973, and V. I. Burenkov [16] in 1974. With the additional assumption that $p > 2 - 1/N$, the theorem was then proved by Hedberg [56], [57] in 1978 and 1981, respectively. Using results by Th. H. Wolff [58], Hedberg finally proved the general case in 1983.

In 1992, Yu. V. Netrusov [89] gave a new proof of the theorem which is valid not only for Sobolev spaces, but also for Besov and Triebel–Lizorkin spaces.

For $p = 1$, the theorem was proved in a special case by A. Carlsson [21] in 1993 and in full generality by Netrusov [90] in 1994.

Hedberg's result has been extended to weighted Sobolev spaces by N. O. Belova [12], [14] in 1994 and 1996, respectively. Belova assumed that the weight w belongs to Muckenhoupt's class of A_p weights.

Below, there follows a short outline of the book. More details can be found in the introduction to each chapter and in the introductions to most sections.

Chapter 1 is of an introductory character. In the first section, we present the notation and conventions that will be used in this book, while in the second, we summarize some facts from the theory of A_p weights. Proofs are given for some of these results.

Chapter 2 deals primarily with weighted Sobolev spaces. In Section 2.1, we prove counterparts for the space $W_w^{m,p}(\Omega)$ of results, such as approximation theorems, extension theorems, and interpolation inequalities, which are well-known

---

[†]This formulation is rather imprecise. What is meant is that the $B_{m-|\alpha|,p}$-quasicontinuous representative of $D^\alpha u$ vanishes $B_{m-|\alpha|,p}$-quasieverywhere on Ω^c.

for non-weighted Sobolev spaces. These results will then be used throughout the text.

In 1938, S. L. Sobolev [97] showed that if $0 < \alpha < N$, $1 < p < N/\alpha$, and the exponent p^* is defined by $1/p^* = 1/p - \alpha/N$, then

$$\left(\int_{\mathbf{R}^N} |\mathcal{I}_\alpha f|^{p^*} \, dx \right)^{1/p^*} \leq C \left(\int_{\mathbf{R}^N} |f|^p \, dx \right)^{1/p}.$$

Here, $\mathcal{I}_\alpha f$ is the Riesz potential of f, i.e., $\mathcal{I}_\alpha f(x) = I_\alpha * f(x)$, where the kernel I_α is given by $I_\alpha(x) = |x|^{-(N-\alpha)}$. The one-dimensional case had been proved ten years earlier by G. H. Hardy and J. E. Littlewood [50]. Using the fact that if u is a smooth function with compact support, then

$$|u(x)| \leq C \int_{\mathbf{R}^N} \frac{|\nabla^m u(y)|}{|x-y|^{N-m}} \, dy$$

(see Landkof [69, p. 45]), it follows that

$$\left(\int_{\mathbf{R}^N} |u|^{p^*} \, dx \right)^{1/p^*} \leq C \left(\int_{\mathbf{R}^N} |\nabla^m u|^p \, dx \right)^{1/p}$$

if $1 < p < N/m$. By approximation with smooth functions, one obtains that if $u \in W^{m,p}(\mathbf{R}^N)$, then actually $u \in L^{p^*}(\mathbf{R}^N)$. This and other embedding theorems are fundamental in Sobolev space theory. In 1974, B. Muckenhoupt and R. L. Wheeden [87] proved a weighted version of Sobolev's inequality. Assuming that $w \in A_{p^*/p'+1}$, one has

$$\left(\int_{\mathbf{R}^N} |\mathcal{I}_\alpha f|^{p^*} w \, dx \right)^{1/p^*} \leq C \left(\int_{\mathbf{R}^N} |f|^p w^{(N-\alpha p)/N} \, dx \right)^{1/p}.$$

As before, this inequality implies that

$$\left(\int_{\mathbf{R}^N} |u|^{p^*} w \, dx \right)^{1/p^*} \leq C \left(\int_{\mathbf{R}^N} |\nabla^m u|^p w^{(N-mp)/N} \, dx \right)^{1/p}$$

for every smooth function u with compact support. G. David and S. Semmes [30] have proved a similar result for the case $p = 1$. Inspired by these last results, we introduce another weighted Sobolev space, denoted by $V_w^{m,p}(\Omega)$, in Section 2.2. The norm of a function u in this space is given by

$$\|u\|_{V_w^{m,p}(\Omega)}^p = \left(\sum_{|\alpha| \leq m} \int_\Omega |D^\alpha u|^p w^{(N-|\alpha|p)/N} \, dx \right)^{1/p}.$$

The integral inequalities obtained in this section have strong similarities with the corresponding inequalities in the non-weighted case.

Some properties of weighted Hausdorff measures are developed in Section 2.3 and used in the following section to prove several weighted isoperimetric inequalities. In some cases, these inequalities generalize earlier results by David and Semmes [30]. Our approach goes back to V. G. Maz'ya [70], who in 1960 showed that the classical isoperimetric inequality:

$$|\Omega|^{(N-1)/N} \le N^{-1}|B_1(0)|^{-1/N}\mathcal{H}^{N-1}(\partial\Omega),$$

where $|\Omega|$ is the N-dimensional measure of the set Ω and $\mathcal{H}^{N-1}(\partial\Omega)$ is the $(N-1)$-dimensional Hausdorff measure of its boundary, is equivalent to the Sobolev type inquality for $p = 1$:

$$\left(\int_\Omega |u|^{N/(N-1)}\,dx\right)^{(N-1)/N} \le N^{-1}|B_1(0)|^{-1/N}\int_\Omega |\nabla u|\,dx,$$

holding for all functions $u \in C_0^\infty(\Omega)$, in the sense that each inequality can be used to prove the other. For completeness, we mention that this inequality is in fact not due to Sobolev, but to E. Gagliardo [44] (1958). Gagliardo's proof did not, however, give the best possible constant. This was obtained by Maz'ya [70] and, independently, by H. Federer and W. H. Fleming [39] in 1960. In Section 2.4, we also prove a weighted version of W. Gustin's "boxing inequality" [49], which will be an important technical tool for us. Gustin's original inequality states that, for every compact set $K \subset \mathbf{R}^N$, there exists a covering $\{B_{r_j}(x_j)\}_{j=1}^\infty$ of K such that

$$\sum_{j=1}^\infty r_j^{N-1} \le C\mathcal{H}^{N-1}(\partial K),$$

where the constant C only depends on N. As consequences of the isoperimetric inequalities, we obtain some weighted Sobolev type inequalities for $p = 1$ that are presented in Section 2.5. The boxing inequality is then employed in Section 2.6 to prove weighted embedding theorems that generalize results by V. G. Maz'ya [73], [75] from the 1970's. These theorems will be of great importance in later sections. Here, we will mention one of Maz'ya's results. Let μ be a positive Radon measure, $1 \le q < \infty$, and $1 \le m < \infty$. Then there exists a constant C such that the inequality

$$\left(\int_{\mathbf{R}^N} |u|^q\,d\mu\right)^{1/q} \le C\int_{\mathbf{R}^N} |\nabla^m u|\,dx$$

holds for every $u \in C_0^\infty(\mathbf{R}^N)$ if and only if

$$\sup_{a\in\mathbf{R}^N,\ r>0} \frac{\mu(B_r(a))^{1/q}}{r^{N-m}} < \infty.$$

By choosing different measures μ, one obtains many of the known embedding theorems. We will for instance show that if $w \in A_1$ and $1 \leq m < N$, then the inequality

$$\int_{\mathbf{R}^N} |u| \, d\mu \leq C \int_{\mathbf{R}^N} |\nabla^m u| w \, dx$$

holds for every $u \in C_0^\infty(\mathbf{R}^N)$ if and only if

$$\sup_{a \in \mathbf{R}^N, \, r > 0} \frac{r^m \mu(B_r(a))}{w(B_r(a))} < \infty,$$

where $w(B_r(a)) = \int_{B_r(a)} w \, dx$.

The theme of Chapter 3 is nonlinear potential theory. In Section 3.1, we prove a weighted norm inequality between the inhomogeneous Riesz potential and the inhomogeneous, fractional maximal function of a positive measure. The corresponding inequality for the homogeneous Riesz potential and the homogeneous, fractional maximal function is a well-known result by B. Muckenhoupt and R. L. Wheeden [87]. This inequality will be used for investigating certain properties of weighted Bessel and Riesz capacities. N. G. Meyers' theory of L^p capacities [77] is then outlined in Section 3.2. This section also contains a generalization of a theorem by Meyers concerning the existence of capacitary measures and capacitary potentials. In Section 3.3, we investigate weighted Bessel and Riesz capacities and, among other things, obtain an elementary proof of D. R. Adams' formula for the weighted Riesz capacity of a ball [5]. Corresponding results for Hausdorff capacities are then established in Section 3.4. The subject of Section 3.5 is variational capacities. We prove that these are comparable to Bessel capacities for $1 < p < \infty$ and to Hausdorff capacities for $p = 1$. The final two sections in Chapter 3 concern weighted thinness. We show that the definition of thinness for the case $1 < p < \infty$, introduced by D. R. Adams in [5], has the Kellogg property. Furthermore, we suggest a definition of thinness for $p = 1$ and prove that the Kellogg property holds.

In Chapter 4, we turn to applications of potential theory to Sobolev spaces. A very useful property of Sobolev functions is the fact that it is possible to speak about their values on sets which have Lebesgue measure zero but positive capacity, for instance, on the boundary of a domain. More precisely, every Sobolev function (viewed as an equivalence class of functions coinciding a.e.) has a representative, which is quasicontinuous with respect to some capacity, i.e., which is continuous on the complement of an open set of arbitrarily small capacity. In particular, this representative is defined outside a set of capacity zero. In the nonlinear case, such results have been established by V. P. Havin and V. G. Maz'ya [53]. It follows from the results in Section 4.1 that if Ω is a Lipschitz domain, then every function $u \in W_w^{m,p}(\Omega)$ has an essentially unique representative, which is quasicontinuous with respect to weighted Bessel capacity, when $1 < p < \infty$, and to weighted Hausdorff capacity, when $p = 1$. In Section 4.2, sufficient conditions for a measure to belong to the dual of $W_w^{m,p}(\Omega)$

are obtained. The corresponding non-weighted results for $p = 1$ are due to
A. Carlsson [20]. These results are then used in Section 4.3 to prove some
weighted Poincaré type inequalities with techniques used earlier by N. G. Meyers [79]. Our inequalities can be thought of as generalizations of the classical
Poincaré inequality:

$$\int_\Omega |u|^p \, dx \leq C \int_\Omega |\nabla u|^p \, dx,$$

which holds for every $u \in C_0^\infty(\Omega)$ if $\Omega \subset \mathbf{R}^N$ is bounded. This inequality is in fact valid as soon as u vanishes on a large enough portion of Ω.
In the strongest possible versions, it suffices to assume that u vanishes on
a subset of Ω with positive capacity. Such inequalities were first studied by
V. G. Maz'ya in [71], [74], and [76]. Other contributions have been made by
A. Carlsson [20], L. I. Hedberg [56], [57], W. P. Ziemer [108], and, in the weighted
case, N. O. Belova [12], [13], [14]. In the final section of the book, we put everything together and prove two results on spectral synthesis, one for the case
$1 < p < \infty$ and one for the case $p = 1$. In both cases, we assume that the set Ω^c
satisfies some extra thickness condition, which, as mentioned before, is fulfilled
if Ω is a bounded Lipschitz domain.

Acknowledgement:
First of all I would like to thank Olli Martio for encouraging me to write this book. I
would also like to express my gratitude to my former supervisor Lars Inge Hedberg,
who introduced me to nonlinear potential theory and supported and advised me during my studies. The book has also profited very much by discussions with Anders
Carlsson and Vladimir G. Maz'ya. I also thank Anders Björn for proofreading and
typographical advice. Finally, I would like to thank almost all of my friends, in particular Gunnar Fogelberg and Hans Åkermark, for never stopping reminding me to finish
the book.

Contents

Chapter 1

Preliminaries

In the first section of this chapter, we introduce the notation and conventions that will be used in this and later chapters. A brief outline of the theory of A_p weights is given in the second section.

1.1. Notation and conventions

Let $\mathbf{R}^N$ denote Euclidian N-space. The norm of a point $x = (x_1, ..., x_N)$ in $\mathbf{R}^N$ is given by $|x| = \left(\sum_{i=1}^{N} x_i^2\right)^{1/2}$. The set $S^{N-1} = \{x \in \mathbf{R}^N \; ; \; |x| = 1\}$ is the unit sphere in $\mathbf{R}^N$. The sets of nonnegative integers and reals are denoted by $\mathbf{Z}_+$ and $\mathbf{R}_+$, respectively. If $E \subset \mathbf{R}^N$, then E°, ∂E, E^c, and $\overline{E}$ stand for the interior, the boundary, the complement, and the closure of E, respectively. The diameter of a set E is denoted $\operatorname{diam} E$ and the distance between two sets A and B is denoted $\operatorname{dist}(A, B)$. The restriction of a function f to a set E is denoted $f|_E$. Let χ_E be the characteristic function of a set E. The set $B_r(a) = \{x \in \mathbf{R}^N \; ; \; |x - a| < r\}$ is an open ball in $\mathbf{R}^N$ with radius r and center a. If $m > 0$, then $mB_r(a) = B_{mr}(a)$. All cubes in $\mathbf{R}^N$ will have their sides parallel to the axes.

We will use the abbreviation "a.e." for "almost everywhere" or "almost every" with respect to Lebesgue measure. Similarly, "measurable" and "locally integrable" always mean Lebesgue measurable and locally integrable with respect to Lebesgue measure, respectively. If Ω is an open subset of $\mathbf{R}^N$, then the set of locally integrable functions on Ω will be denoted by $L^1_{\mathrm{loc}}(\Omega)$. The Lebesgue measure of a measurable subset E of $\mathbf{R}^N$ is denoted $|E|$. The mean value over a set E of a locally integrable function f on $\mathbf{R}^N$ is

$$\fint_E f \, dx = \frac{1}{|E|} \int_E f \, dx.$$

We will use a similar notation for mean values with respect to arbitrary measures. If $1 \leq p \leq \infty$, then p' is the conjugate exponent to p given by $1/p + 1/p' = 1$ with the usual conventions when $p = 1$ or $p = \infty$. The sets of Radon measures and positive Radon measures,[†] concentrated to a set E, are denoted $\mathcal{M}(E)$ and $\mathcal{M}^+(E)$, respectively. Let $\mu|_E$ be the restriction of a measure μ to a μ-measurable set E.

Let $\alpha = (\alpha_1, \ldots, \alpha_N) \in \mathbf{Z}_+^N$ be a multi-index. Then $|\alpha| = \alpha_1 + \ldots + \alpha_N$, $\alpha! = \alpha_1! \cdot \ldots \cdot \alpha_N!$, $x^\alpha = x_1^{\alpha_1} \cdot \ldots \cdot x_N^{\alpha_N}$ for $x \in \mathbf{R}^N$, and

$$D^\alpha = \frac{\partial^{\alpha_1}}{\partial x_1^{\alpha_1}} \cdots \frac{\partial^{\alpha_N}}{\partial x_N^{\alpha_N}}.$$

We also let $D_i = \partial / \partial x_i$. If α and β are two multi-indices, we shall write $\beta \leq \alpha$ provided $\beta_i \leq \alpha_i$ for $i = 1, \ldots, N$.

If Ω is an open subset of $\mathbf{R}^N$ and $0 \leq k \leq \infty$, then $C^k(\Omega)$ is the set of k times continuously differentiable functions on Ω and $C_0^k(\Omega)$ is the set of k times continuously differentiable functions, having compact support in Ω. The support of a function u will be denoted $\operatorname{supp} u$. The gradient of u is $\nabla u = (\partial_1 u, \ldots, \partial_N u)$. We also use the notation

$$|\nabla^k u| = \sum_{|\alpha| = k} |D^\alpha u|$$

when $k \geq 2$ is an integer. The space of polynomials in N variables of degree $\leq m$ is denoted $\mathcal{P}_m$. The Schwartz class of rapidly decreasing functions on $\mathbf{R}^N$ is denoted $\mathcal{S}$.

Within a proof of, say, a theorem, the letter C (and occasionally other letters) will be used to denote a generic constant, that only depends on the parameters in the statement of the theorem. The value of C may thus change from one occurrence to another. Two quantities A and B are said to be "equivalent" or "comparable" if there exists two constants C_1 and C_2 so that $C_1 A \leq B \leq C_2 A$. The symbol $\square$ is used to mark the end of a proof.

1.2. Basic results concerning weights

In this section, we review some properties of weights and, in particular, A_p weights, that will be used throughout this book. Complete expositions can be found in the monographs by J. García–Cuerva and J. L. Rubio de Francia [45] and A. Torchinsky [102].

1.2.1. General weights

By a *weight*, we shall mean a locally integrable function w on $\mathbf{R}^N$ such that $w(x) > 0$ for a.e. $x \in \mathbf{R}^N$. Every weight w gives rise to a measure on the mea-

---

[†]By a Radon measure we here mean a Borel measure on $\mathbf{R}^N$ with the additional properties that every subset of $\mathbf{R}^N$ is contained in Borel set with equal measure and that every compact set has finite measure.

surable subsets of $\mathbf{R}^N$ through integration. This measure will also be denoted by w. Thus, $w(E) = \int_E w\,dx$ for measurable sets $E \subset \mathbf{R}^N$.

Definition 1.2.1. Let w be a weight, and let $\Omega \subset \mathbf{R}^N$ be open. For $0 < p < \infty$, we define $L^p_w(\Omega)$ as the set of measurable functions f on Ω such that

$$\|f\|_{L^p_w(\Omega)} = \left(\int_\Omega |f|^p w\,dx \right)^{1/p} < \infty.$$

We also define wk-$L^1_w(\Omega)$ as the set of measurable functions f on Ω satisfying

$$\|f\|_{\text{wk-}L^1_w(\Omega)} = \sup_{\lambda > 0} \lambda w(\{x \in \Omega\,;\, |f(x)| > \lambda\}) < \infty.$$

If ν is a positive Borel measure on an open set Ω, we shall more generally denote by $L^p_\nu(\Omega)$, $0 < p < \infty$, the set of ν-measurable functions f on Ω for which

$$\left(\int_\Omega |f|^p \, d\nu \right)^{1/p} < \infty.$$

We now determine conditions on the weight w that guarantee that functions in $L^p_w(\Omega)$ are locally integrable on Ω. Let $1 \le p < \infty$ and let w be a weight such that $w^{-1/(p-1)}$ is locally integrable, when $p > 1$, and such that w locally is bounded from below away from 0, when $p = 1$, i.e., such that

$$\operatorname*{ess\,sup}_{x \in B} \frac{1}{w(x)} < \infty$$

for every ball B. Suppose that $f \in L^p_w(\Omega)$, and let $B \subset \Omega$ be a ball. If $1 < p < \infty$, then, by Hölder's inequality,

$$\int_B |f|\,dx \le \left(\int_B w^{-1/(p-1)}\,dx \right)^{1/p'} \left(\int_B |f|^p w\,dx \right)^{1/p}, \qquad (1.2.1)$$

and if $p = 1$,

$$\int_B |f|\,dx \le \operatorname*{ess\,sup}_{x \in B} \frac{1}{w(x)} \int_B |f| w\,dx. \qquad (1.2.2)$$

It follows that $L^p_w(\Omega) \subset L^1_{\text{loc}}(\Omega)$ and that convergence in $L^p_w(\Omega)$ implies local convergence in $L^1(\Omega)$. If Ω is bounded, one obtains in the same way that $L^p_w(\Omega)$ is continuously embedded in $L^1(\Omega)$.

1.2.2. A_p weights

The class of A_p weights was introduced by B. Muckenhoupt in [85], where he showed that the A_p weights are precisely those weights w for which the Hardy–Littlewood maximal operator is bounded from $L^p_w(\mathbf{R}^N)$ to $L^p_w(\mathbf{R}^N)$, when $1 < p < \infty$, and from $L_w^1(\mathbf{R}^N)$ to wk-$L_w^1(\mathbf{R}^N)$, when $p = 1$. Here, we define the *Hardy–Littlewood maximal function*, Mf, for a locally integrable function f on $\mathbf{R}^N$ by

$$Mf(x) = \sup_{r>0} \fint_{B_r(x)} |f(y)|\, dy.$$

The corresponding operator, which takes f to Mf, is denoted by M.

We begin by defining the class of A_p weights.

Definition 1.2.2. Let $1 \le p < \infty$. A weight w is said to be an A_p *weight*, if there exists a positive constant A such that, for every ball $B \subset \mathbf{R}^N$,

$$\left(\fint_B w\, dx\right)\left(\fint_B w^{-1/(p-1)}\, dx\right)^{p-1} \le A, \tag{1.2.3}$$

if $p > 1$, or

$$\left(\fint_B w\, dx\right) \operatorname*{ess\,sup}_{x \in B} \frac{1}{w(x)} \le A, \tag{1.2.4}$$

if $p = 1$. The infimum over all such constants A is called the A_p *constant* of w. We denote by A_p, $1 \le p < \infty$, the set of all A_p weights.

We will refer to (1.2.3) and (1.2.4) as the A_p and the A_1 condition, respectively. Muckenhoupt's theorem is now the following [85, p. 209, p. 222].

Theorem 1.2.3. *Suppose that $w \in A_p$, where $1 < p < \infty$. Then the Hardy–Littlewood maximal operator M is bounded on $L_w^p(\mathbf{R}^N)$, that is, there exists a positive constant C such that*

$$\int_{\mathbf{R}^N} (Mf)^p w\, dx \le C \int_{\mathbf{R}^N} |f|^p w\, dx \tag{1.2.5}$$

for every $f \in L_w^p(\mathbf{R}^N)$. The constant C depends only on N, p, and the A_p constant of w. If $w \in A_1$, then M is bounded from $L_w^1(\mathbf{R}^N)$ to wk-$L_w^1(\mathbf{R}^N)$. In other words,

$$w(\{x \in \mathbf{R}^N \; ; \; Mf(x) > \lambda\}) \le \frac{C}{\lambda} \int_{\mathbf{R}^N} |f| w\, dx \tag{1.2.6}$$

for every $f \in L_w^1(\mathbf{R}^N)$ and every $\lambda > 0$, with a constant C that only depends on N and the A_1 constant of w. Conversely, if (1.2.5) holds for every $f \in L_w^p(\mathbf{R}^N)$, then $w \in A_p$, and if (1.2.6) holds for every $f \in L_w^1(\mathbf{R}^N)$, then $w \in A_1$.

In Section 1.2.5, we shall prove that (1.2.5) holds if $w \in A_p$. This is the only part of Theorem 1.2.3 that will be used in the rest of this book.

Remark 1.2.4. Below we list some simple, but useful properties of A_p weights.

1. If $w \in A_p$, $1 \leq p < \infty$, then since $w^{-1/(p-1)}$ is locally integrable, when $p > 1$, and $1/w$ is locally bounded, when $p = 1$, we have $L^p_w(\Omega) \subset L^1_{\text{loc}}(\Omega)$ for every domain Ω. Moreover, if A is the A_p constant of w, then by the A_p condition, the right-hand sides of (1.2.1) and (1.2.2) do not exceed

$$A^{1/p}|B| \left(\frac{1}{w(B)} \int_B |f|^p \, w \, dx \right)^{1/p} .$$

2. Note that if w is a weight, then, by writing $1 = w^{1/p} w^{-1/p}$, Hölder's inequality implies that, for every ball B,

$$1 \leq \left(\fint_B w \, dx \right) \left(\fint_B w^{-1/(p-1)} \, dx \right)^{p-1} ,$$

when $p > 1$, and similarly for the expression that gives the A_1 condition. It follows that if $w \in A_p$, then the A_p constant of w is ≥ 1.

3. It also follows from Hölder's inequality that if $1 \leq p < q < \infty$, then $A_p \subset A_q$, and the A_q constant of a weight $w \in A_p$ equals the A_p constant of w.

4. If $w \in A_p$, where $1 < p < \infty$, then $w^{-1/(p-1)} \in A_{p'}$, and conversely. When p is fixed, we shall sometimes denote the weight $w^{-1/(p-1)}$ by w'.

5. The A_p condition is invariant under translations and dilations, i.e., if $w \in A_p$, then the weights $x \mapsto w(x + a)$ and $x \mapsto w(\delta x)$, where $a \in \mathbf{R}^N$ and $\delta > 0$ are fixed, both belong to A_p with the same A_p constants as w.

6. As it sometimes is more convenient to work with cubes than balls, it is useful to notice that if one replaces the balls in the definition of A_p with cubes, one gets the same class of weights and the different "A_p constants" are comparable.

7. It is not so difficult to see that a weight w belongs to A_1 if and only if $Mw(x) \leq Aw(x)$ a.e.

8. It follows that if $w \in A_1$, then there is a constant C such that

$$w(x) \geq \frac{C}{(1+|x|)^N}$$

for a.e. $x \in \mathbf{R}^N$. In fact, if $x \in \mathbf{R}^N$ and $r = 2\max\{1, |x|\}$, then

$$\frac{1}{r^N} \int_{B_r(x)} w \, dy \geq \frac{1}{2^N} \frac{1}{(1+|x|)^N} \int_{B_1(0)} w \, dy,$$

so $Mw(x) \geq C(1 + |x|)^{-N}$ a.e. This is of course the same argument that shows that the maximal function of a locally integrable function is never integrable on $\mathbf{R}^N$ (unless the function is identically zero).

Example 1.2.5. It might be useful for the reader to have some examples of A_p weights in mind when reading the rest of this book.

1. If w is a weight and there exist two positive constants C and D such that $C \leq w(x) \leq D$ for a.e. $x \in \mathbf{R}^N$, then obviously $w \in A_p$ for $1 \leq p < \infty$.

2. Suppose that $w(x) = |x|^\eta$. Then $w \in A_1$, if $-N < \eta \leq 0$, and $w \in A_p$, $1 < p < \infty$, if $-N < \eta < N(p-1)$ (see Torchinsky [102, p. 229, p. 236]).[†]

3. There is a connection between A_p and BMO, the class of functions with bounded mean oscillation. In fact, if w is a weight, then $\log w \in$ BMO if and only if $w^\eta \in A_2$ for some $\eta > 0$ (see Torchinsky [102, p. 240]).

4. Let μ be a positive Borel measure such that the maximal function $M\mu$ is not identically ∞. Then $(M\mu)^\varepsilon \in A_1$ for $0 \leq \varepsilon < 1$. Conversely, every A_1 weight w can be factored $w = b(Mf)^\varepsilon$, where $0 < C \leq b(x) \leq D < \infty$ for a.e. $x \in \mathbf{R}^N$, $f \in L^1_{\text{loc}}(\mathbf{R}^N)$, $0 \leq \varepsilon < 1$, and $(Mf)^\varepsilon$ is finite a.e. This is a result by R. R. Coifman and R. Rochberg [29]; see also Torchinsky [102, pp. 229–230].

1.2.3. Doubling weights

We will often use the fact that A_p weights are doubling.

Definition 1.2.6. A weight w is said to be *doubling*, if there exists a positive constant C such that

$$w(2B) \leq Cw(B) \tag{1.2.7}$$

for every ball $B \subset \mathbf{R}^N$. The infimum over all constants C, for which (1.2.7) holds, is called the *doubling constant* of w.

It follows directly from the A_p condition and Hölder's inequality that an A_p weight has the following *strong doubling property*. In particular, every A_p weight is doubling.

Proposition 1.2.7. *Let $w \in A_p$, where $1 \leq p < \infty$, and let E be a measurable subset of a ball B. Then*

$$w(B) \leq A \left(\frac{|B|}{|E|} \right)^p w(E),$$

where A is the A_p constant of w.

[†]Note that $|x|$ is the distance from x to the origin in $\mathbf{R}^N$. One expects that suitable powers of distances to linear subspaces of $\mathbf{R}^N$ and to submanifolds of $\mathbf{R}^N$ should be A_p weights. No general results are, however, known to the author.

Proof. We have

$$|E| \leq \left(\int_E w \, dx \right)^{1/p} \left(\int_E w^{-1/(p-1)} \, dx \right)^{(p-1)/p}$$

$$\leq w(E)^{1/p} |B|^{(p-1)/p} \left(\fint_B w^{-1/(p-1)} \, dx \right)^{(p-1)/p}$$

$$\leq A^{1/p} w(E)^{1/p} |B|^{(p-1)/p} \left(\fint_B w \, dx \right)^{-1/p}$$

$$= A^{1/p} \left(\frac{w(E)}{w(B)} \right)^{1/p} |B|. \quad \square$$

1.2.4. A_∞ weights

Another important class of weights is the class of A_∞ weights, introduced by C. Fefferman (see Gundy–Wheeden [48, p. 108]). The following definition of A_∞, just one of several equivalent ones, suits our purposes best.

Definition 1.2.8. We say that a weight w is an A_∞ *weight*, if there exist two positive constants C and δ such that

$$w(Q) \geq C \left(\frac{|Q|}{|E|} \right)^\delta w(E)$$

for every cube Q and every measurable subset E of Q. The constants C and δ are called A_∞ *constants* of w and the set of A_∞ weights is (of course) denoted A_∞.

The relationship between A_p and A_∞ is clarified by the two theorems below, due to Muckenhoupt [85, p. 214], [86, p. 104]. Together they show that

$$A_\infty = \bigcup_{1 \leq p < \infty} A_p.$$

Theorem 1.2.9. *If $w \in A_p$, $1 \leq p < \infty$, then $w \in A_\infty$ with A_∞ constants that only depend on N and the A_p constant of w.*

Theorem 1.2.10. *If $w \in A_\infty$, then $w \in A_p$ for some p, $1 < p < \infty$, and the A_p constant of w is majorized by a constant that only depends on N and the A_∞ constants of w.*

Remark 1.2.11. A consequence of Theorem 1.2.9 and the defining condition for A_∞ is the fact that $\int_{\mathbf{R}^N} w \, dx = \infty$ for every weight $w \in A_p$.

We now set out to prove these theorems, using an approach due to R. Coifman and C. Fefferman [28], and begin by proving a reverse Hölder inequality.

Lemma 1.2.12. *Suppose that μ_1 is a positive Radon measure on $\mathbf{R}^N$ satisfying $\mu_1(\mathbf{R}^N) = \infty$, and let the function w be locally integrable on $\mathbf{R}^N$ with respect to μ_1. Suppose that there are two constants α and β such that*

$$\mu_1(\{x \in Q \,;\, w(x) > \beta\overline{w}(Q)\}) \geq \alpha\mu_1(Q) \qquad (1.2.8)$$

for every cube Q, where

$$\overline{w}(Q) = \fint_Q w \, d\mu_1.$$

Then there are constants C and η such that the inequality

$$\left(\fint_Q w^{1+\eta} \, d\mu_1\right)^{1/(1+\eta)} \leq C \fint_Q w \, d\mu_1 \qquad (1.2.9)$$

holds for every cube Q.

Proof. Let Q be a cube, and let $\lambda > 0$ be arbitrary. According to the Calderón–Zygmund lemma [99], there is a family $\{Q_i\}$ of pairwise disjoint subcubes of Q with the properties

$$w(x) \leq \lambda \quad \text{for } \mu_1\text{-a.e. } x \in Q \setminus \bigcup_i Q_i$$

and

$$\lambda < \fint_{Q_i} w \, d\mu_1 \leq 2^N \lambda.$$

Set $d\mu_2(x) = w(x) \, d\mu_1(x)$. It follows that

$$\mu_2(\{x \in Q \,;\, w(x) > \lambda\}) \leq \sum_i \mu_2(Q_i) \leq 2^N \lambda \sum_i \mu_1(Q_i)$$

$$\leq \frac{2^N \lambda}{\alpha} \sum_i \mu_1(\{x \in Q_i \,;\, w(x) > \beta\overline{w}(Q_i)\})$$

$$\leq \frac{2^N \lambda}{\alpha} \sum_i \mu_1(\{x \in Q_i \,;\, w(x) > \beta\lambda\})$$

$$\leq \frac{2^N \lambda}{\alpha} \mu_1(\{x \in Q \,;\, w(x) > \beta\lambda\}). \qquad (1.2.10)$$

Now multiply (1.2.10) with $\lambda^{\eta-1}$, where λ is as yet to be specified, and integrate to obtain

$$\int_{\overline{w}(Q)}^{\infty} \lambda^{\eta-1} \mu_2(\{x \in Q \,;\, w(x) > \lambda\}) \, d\lambda$$

$$\leq C \int_0^{\infty} \lambda^{\eta} \mu_1(\{x \in Q \,;\, w(x) > \beta\lambda\}) \, d\lambda = \frac{C}{1+\eta} \int_Q w^{1+\eta} \, d\mu_1.$$

An application of Fubini's theorem shows that

$$\int_{\overline{w}(Q)}^{\infty} \lambda^{\eta-1} \mu_2(\{x \in Q \,;\, w(x) > \lambda\}) \, d\lambda$$

$$= \int_{\{x \in Q; w(x) > \overline{w}(Q)\}} w(x) \left(\int_{\overline{w}(Q)}^{w(x)} \lambda^{\eta-1} \, d\lambda \right) d\mu_1(x)$$

$$= \int_{\{x \in Q; w(x) > \overline{w}(Q)\}} w(x) \left(\frac{w(x)^{\eta}}{\eta} - \frac{\overline{w}(Q)^{\eta}}{\eta} \right) d\mu_1(x)$$

$$\geq \frac{1}{\eta} \int_Q w(x)^{1+\eta} \, d\mu_1(x) - \frac{2}{\eta} \overline{w}(Q)^{1+\eta} \mu_1(Q).$$

Putting everything together, we see that

$$\left(\frac{1}{\eta} - \frac{C}{1+\eta} \right) \left(\fint_Q w^{1+\eta} \, d\mu_1 \right)^{1/(1+\eta)} \leq \frac{2}{\eta} \overline{w}(Q),$$

and (1.2.9) follows by choosing η small enough. $\square$

Definition 1.2.13. Let μ_1 and μ_2 be two positive measures on $\mathbf{R}^N$. We say that μ_1 is *comparable* to μ_2, if there exist two constants α and β such that if E is a measurable subset of a cube Q, then

$$\frac{\mu_2(E)}{\mu_2(Q)} \leq \beta \quad \text{implies} \quad \frac{\mu_1(E)}{\mu_1(Q)} \leq \alpha.$$

Lemma 1.2.14. *Suppose that μ_1 is a positive Radon measure on $\mathbf{R}^N$ satisfying $\mu_1(\mathbf{R}^N) = \infty$, and let the function w be locally integrable on $\mathbf{R}^N$ with respect to μ_1. Set $d\mu_2(x) = w(x) \, d\mu_1(x)$. Suppose that μ_1 is comparable to μ_2. Then the reverse Hölder inequality in Lemma 1.2.12 holds.*

Proof. It suffices to show that the inequality (1.2.8) holds. Let

$$E = \{x \in Q \,;\, w(x) > \beta \overline{w}(Q)\},$$

and set $E' = Q \setminus E$. Then because $w(x) \leq \beta \overline{w}(Q)$ on E',

$$\frac{\mu_2(E')}{\mu_2(Q)} = \frac{1}{\mu_2(Q)} \int_{E'} w \, d\mu_1 \leq \frac{1}{\mu_2(Q)} \frac{\mu_2(Q)}{\mu_1(Q)} \beta \mu_1(E') \leq \beta,$$

so it follows from the assumption that $\mu_1(E') \leq \alpha \mu_1(Q)$, whence

$$\mu_1(E) \geq (1-\alpha) \mu_1(Q). \quad \square$$

Lemma 1.2.15. *Let $1 < p < \infty$. If $w \in A_p$, with A_p constant A, then there is a constant C, that only depends on N, such that the inequality*

$$\left| \left\{ x \in Q \,;\, w(x) > \beta \fint_Q w \, dy \right\} \right| \geq (1 - (CA\beta)^{1/(p-1)}) |Q| \qquad (1.2.11)$$

holds for every number β satisfying $0 < \beta \leq 1/(CA)$ and every cube $Q \subset \mathbf{R}^N$.

Remark 1.2.16. The same inequality holds for balls with the only difference that CA is replaced by A. This fact will be used in Section 2.2 for the proof of Lemma 2.2.17.

Proof. Let E be the set in the left-hand side of (1.2.11), and set $E' = Q \setminus E$. By the A_p condition and Remark 1.2.4.6, we then have

$$\frac{1}{\beta}\left(\frac{|E'|}{|Q|}\right)^{p-1} = \overline{w}(Q)\left(\frac{1}{|Q|}\int_{E'}(\beta\overline{w}(Q))^{-1/(p-1)}dx\right)^{p-1}$$
$$\leq \overline{w}(Q)\left(\fint_Q w(x)^{-1/(p-1)}dx\right)^{p-1} \leq CA,$$

from which (1.2.11) easily follows. □

Proof of Theorem 1.2.9. Lemma 1.2.12 and Lemma 1.2.15 together show that one can find constants C and $\eta > 0$ so that the reverse Hölder inequality

$$\left(\fint_Q w^{1+\eta}\,dx\right)^{1/(1+\eta)} \leq C\fint_Q w\,dx$$

holds for every cube Q. If $E \subset Q$ is measurable, then first Hölder's inequality and then the reverse inequality imply

$$w(E) \leq |E|^{\eta/(1+\eta)}\left(\int_Q w^{1+\eta}\,dx\right)^{1/(1+\eta)} \leq C|Q|^{1/(1+\eta)}|E|^{\eta/(1+\eta)}\fint_Q w\,dx$$
$$= C\left(\frac{|E|}{|Q|}\right)^{\eta/(1+\eta)}w(Q).$$

Thus, $w \in A_\infty$ with $\delta = \eta/(1+\eta)$. □

Proof of Theorem 1.2.10. Take

$$d\mu_1(x) = w(x)\,dx \quad\text{and}\quad d\mu_2(x) = dx = w(x)^{-1}\,d\mu_1$$

in Lemma 1.2.12. By the A_∞ condition, μ_1 is comparable to μ_2. Thus, there holds

$$\left(\fint_Q\left(\frac{1}{w}\right)^{1+\eta}\,d\mu_1\right)^{1/(1+\eta)} \leq C\fint_Q\frac{1}{w}\,d\mu_1$$

for every cube Q. But, after some rearrangements, this condition states precisely that $w \in A_p$, where $p = 1 + 1/\eta$. □

1.2.5. Proof of Muckenhoupt's maximal theorem

The reverse Hölder inequality in the proof of Theorem 1.2.9 may be used to prove the so called "open-end property" of A_p.

Corollary 1.2.17. *Suppose that $w \in A_p$ for some p, $1 < p < \infty$. Then there exists a number q, $1 < q < p$, such that $w \in A_q$.*

Proof. By Remark 1.2.4.4, the weight $w^{-1/(p-1)} \in A_{p'}$. It thus follows from the proof of Theorem 1.2.9 that there exist positive constants C and η such that the reverse Hölder inequality

$$\left(\fint_Q w^{-(1+\eta)/(p-1)} \, dx \right)^{1/(1+\eta)} \leq C \fint_Q w^{-1/(p-1)} \, dx. \qquad (1.2.12)$$

Now let q be determined by $(1+\eta)/(p-1) = 1/(q-1)$, Then $1 < q < p$, and by (1.2.12) and the A_p condition,

$$\left(\fint_Q w^{-1/(q-1)} \, dx \right)^{q-1} = \left(\fint_Q w^{-(1+\eta)/(p-1)} \, dx \right)^{(p-1)/(1+\eta)}$$

$$\leq C \left(\fint_Q w^{-1/(p-1)} \, dx \right)^{p-1} \leq C \left(\fint_Q w \, dx \right)^{-1},$$

which shows that $w \in A_q$. $\square$

We now come to the proof of Theorem 1.2.3. As mentioned above, we only prove that the inequality (1.2.5) holds if $w \in A_p$. The proof is due to Coifman and Fefferman [28, pp. 242–243].

Proof of Theorem 1.2.3. Let us introduce a weighted maximal function:

$$M_w f(x) = \sup_{r>0} \frac{1}{w(B_r(x))} \int_{B_r(x)} |f(y)| w(y) \, dy, \qquad x \in \mathbf{R}^N.$$

The standard proof of the Hardy–Littlewood–Wiener maximal theorem in, e.g., Stein [99, pp. 6–7] (with obvious changes) shows that if $g \in L_w^s(\mathbf{R}^N)$, $1 < s < \infty$, then

$$\int_{\mathbf{R}^N} (M_w g)^s w \, dx \leq C \int_{\mathbf{R}^N} |g|^s w \, dx.^\dagger \qquad (1.2.13)$$

Let B be an arbitrary ball, and suppose that $w \in A_q$ for $1 < q < p$. It follows from Hölder's inequality and Remark 1.2.4.1 that

$$\fint_B |f| \, dy \leq C \left(\frac{1}{w(B)} \int_B |f|^q w \, dy \right)^{1/q},$$

whence $Mf(x) \leq C(M_w|f|^q(x))^{1/q}$ for every $x \in \mathbf{R}^N$. The inequality (1.2.5) now follows from (1.2.13) with $g = |f|^q$ and $s = p/q$:

$$\int_{\mathbf{R}^N} (Mf)^p w \, dx \leq C \int_{\mathbf{R}^N} (M_w|f|^q)^{p/q} w \, dx \leq C \int_{\mathbf{R}^N} |f|^p w \, dx. \quad \square$$

---

†The proof of this inequality uses only the fact that w is doubling.

1.2.6. Boundedness of singular integrals

The theorem below, by Coifman and Fefferman [28, p. 244], shows that the standard singular integral operators are bounded on A_p weighted L^p spaces. We omit the proof since this theorem will not be used later on.

Theorem 1.2.18. *Suppose that the kernel* $K : \mathbf{R}^N \setminus \{0\} \to \mathbf{R}$ *satisfies*

$$\|\widehat{K}\|_{L^\infty(\mathbf{R}^N)} \leq A,$$

where $\widehat{K}$ *is the fourier transform of* K,

$$|K(x)| \leq \frac{A}{|x|^N}, \qquad \text{for every } x \in \mathbf{R}^N \setminus \{0\},$$

and

$$|K(x) - K(x-y)| \leq \frac{A|y|}{|x|^{N+1}},$$

if $x, y \in \mathbf{R}^N$ *and* $|x| > 2|y|$. *Let* $w \in A_\infty$, *and define for* $f \in L_w^p(\mathbf{R}^N)$, *where* $0 < p < \infty$, *and* $\varepsilon > 0$,

$$T_\varepsilon f(x) = \int_{|x-y| \geq \varepsilon} K(x-y) f(y) \, dy$$

and

$$T^* f(x) = \sup_{\varepsilon > 0} |T_\varepsilon f(x)|.$$

Then

$$\|T^* f\|_{L_w^p(\mathbf{R}^N)} \leq C \|Mf\|_{L_w^p(\mathbf{R}^N)}$$

for every $f \in L_w^p(\mathbf{R}^N)$, *where the constant* C *only depends on* A, N, p, *and the* A_∞ *constants of* w. *Moreover, if* $w \in A_p$, $1 < p < \infty$, *and* $Tf = K * f$, *then*

$$\|Tf\|_{L_w^p(\mathbf{R}^N)} \leq C \|f\|_{L_w^p(\mathbf{R}^N)}$$

for every $f \in L_w^p(\mathbf{R}^N)$, *with a constant* C *that only depends on* A, N, p, *and the* A_p *constant of* w.

1.2.7. Two theorems by Muckenhoupt and Wheeden

We close this section with two theorems by Muckenhoupt and R. L. Wheeden [87, pp. 262, 267], that deal with Riesz potentials and fractional maximal functions. The proofs of the theorems will be omitted.

Definition 1.2.19. Let $0 < \alpha < N$. The *Riesz kernel* is the function $I_\alpha(x) = |x|^{-(N-\alpha)}$. If μ is a measure on $\mathbf{R}^N$, then the *Riesz potential* of μ, $\mathcal{I}_\alpha\mu$, is defined by $\mathcal{I}_\alpha\mu = I_\alpha * \mu$. Similarly, for $0 < \rho < \infty$, the *inhomogeneous Riesz kernel* is

$$I_{\alpha,\rho}(x) = \begin{cases} |x|^{-(N-\alpha)}, & \text{if } |x| < \rho, \\ 0, & \text{if } |x| \geq \rho, \end{cases}$$

and the *inhomogeneous Riesz potential* $\mathcal{I}_{\alpha,\rho}\mu$ of μ is defined by $\mathcal{I}_{\alpha,\rho}\mu = I_{\alpha,\rho} * \mu$.

If $d\mu = f\,dx$, where f is a measurable function on $\mathbf{R}^N$, we will denote the potentials of μ by $\mathcal{I}_\alpha f$ and $\mathcal{I}_{\alpha,\rho} f$, respectively.

Definition 1.2.20. Let μ be a positive measure on $\mathbf{R}^N$, and let $0 < \alpha < N$. The *fractional maximal function* of μ, $M_\alpha\mu$, is defined by

$$M_\alpha\mu(x) = \sup_{0<r<\infty} \frac{1}{r^{N-\alpha}} \int_{B_r(x)} d\mu(y).$$

If $0 < \rho < \infty$, then the *inhomogeneous maximal function* $M_{\alpha,\rho}\mu$ of μ is similarly defined as

$$M_{\alpha,\rho}\mu(x) = \sup_{0<r<\rho} \frac{1}{r^{N-\alpha}} \int_{B_r(x)} d\mu(y).$$

If $d\mu = |f|\,dx$, we denote these maximal functions by $M_\alpha f$ and $M_{\alpha,\rho} f$, respectively. Note that $M_0 f$ is a constant times Mf.

Theorem 1.2.21 shows that the Riesz potential of a function is controlled in norm by the corresponding fractional maximal function. An inhomogeneous version of this theorem will be proved in Section 3.1. Muckenhoupt and Wheeden used this theorem together with a maximal inequality for $M_\alpha\mu$ to establish Theorem 1.2.22, a weighted Sobolev type inequality for Riesz potentials.

Theorem 1.2.21. *Suppose that $w \in A_\infty$, and let $0 < q < \infty$ and $0 < \alpha < N$. Then there is a constant C, that only depends on N, q, and the A_∞ constants of w, such that*

$$\int_{\mathbf{R}^N} |\mathcal{I}_\alpha f|^q w\,dx \leq C \int_{\mathbf{R}^N} (M_\alpha f)^q w\,dx$$

for every measurable function f.

Theorem 1.2.22. *Let v be a weight, and let $\alpha > 0$ and $1 < p < N/\alpha$. Set $1/p^* = 1/p - \alpha/N$. Then*

$$\left(\int_{\mathbf{R}^N} |\mathcal{I}_\alpha f|^{p^*} v^{p^*}\,dx \right)^{1/p^*} \leq C \left(\int_{\mathbf{R}^N} |f|^p v^p\,dx \right)^{1/p}$$

for every $f \in L_{v^p}^p(\mathbf{R}^N)$ if and only if

$$\sup_B \left(\fint_B v^{p^*}\,dx \right) \left(\fint_B v^{-p'}\,dx \right)^{p^*/p'} < \infty,$$

where the supremum is taken over all balls $B \subset \mathbf{R}^N$.

Remark 1.2.23. The following observation is taken from the proof of Theorem 1.2.22 [87, p. 268]. Set $w = v^{p^*}$. Then $v^p = w^{(N-\alpha p)/N}$ and $v^{-p'} = w^{p'/p^*}$. Hence, if we let $p^*/p' = q - 1$, we have

$$\left(\int_{\mathbf{R}^N} |I_\alpha * f|^{p^*} w \, dx \right)^{1/p^*} \le C \left(\int_{\mathbf{R}^N} |f|^p w^{(N-\alpha p)/N} \, dx \right)^{1/p} \qquad (1.2.14)$$

for every function f if and only if

$$\sup_B \left(\fint_B w \, dx \right) \left(\fint_B w^{-1/(q-1)} \, dx \right)^{q-1} < \infty,$$

i.e., if and only if $w \in A_q$. Note that $q > p$, so by Remark 1.2.11.3, the inequality (1.2.14) holds, e.g., if $w \in A_p$. If $w \in A_q$, then the constant in (1.2.14) depends only on α, N, p, and the A_q constant of w.

Chapter 2

Sobolev spaces

The present chapter explores some properties of weighted Sobolev spaces. As mentioned already in the preface, we will consider two types of Sobolev spaces, namely the spaces $W_w^{m,p}(\Omega)$ and $V_w^{m,p}(\Omega)$. The norm of a function u in these spaces is

$$\|u\|_{W_w^{m,p}(\Omega)} = \left(\sum_{|\alpha| \leq m} \int_\Omega |D^\alpha u|^p w \, dx \right)^{1/p}$$

and

$$\|u\|_{V_w^{m,p}(\Omega)} = \left(\sum_{|\alpha| \leq m} \int_\Omega |D^\alpha u|^p w^{(N-|\alpha|p)/N} \, dx \right)^{1/p},$$

respectively. In the first two sections, we prove results concerning approximation and extension. We also prove several norm inequalities. The inequalities obtained for the space $V_w^{m,p}(\Omega)$ bear a close resemblance to the corresponding inequalities for $w = 1$.

In Section 2.3, some properties for weighted Hausdorff measures are established. Some of the results in Section 2.2 are then used in the following section to prove an extension of a two-weighted isoperimetric inequality due to G. David and S. Semmes [30]. This inequality in turn implies a single weighted isoperimetric inequality. We also obtain isoperimetric inequalities that involve weighted $(N-1)$-dimensional, lower Minkowski content and weighted Hausdorff measures. Another result in Section 2.4 is a generalization of W. Gustin's boxing inequality [49], that will be employed in Section 2.6 for proving certain embedding theorems.

In the non-weighted case, it is well-known that the isoperimetric inequality implies the Sobolev type inequality for the space $\overset{\circ}{W}^{m,1}(\Omega)$. It should therefore not come as a surprise to the reader that we are able to obtain weighted Sobolev type inequalities in Section 2.5 with the aid of the results in Section 2.4.

In Section 2.6, necessary and sufficient conditions for the inequalities

$$\int_{\mathbf{R}^N} |u|\, d\mu \le C \int_{\mathbf{R}^N} |\nabla^m u| w\, dx$$

and

$$\int_{\mathbf{R}^N} |u|\, d\mu \le C \|u\|_{W_w^{m,1}(\mathbf{R}^N)}$$

(as well as other inequalities) to hold for every function $u \in C_0^\infty(\mathbf{R}^N)$ are obtained, where μ is a given positive Radon measure. The corresponding results in the non-weighted case are due to V. G. Maz'ya [73], [75]. These inequalities will be used later in Section 3.5 to prove that the Hausdorff capacities introduced in Section 3.4 are equivalent to variational capacities, in Section 4.1 to prove the existence of quasicontinuous representatives of functions belonging to $W_w^{m,1}(\Omega)$, and in Section 4.2 to give a sufficient condition for a measure to belong to the dual of $W_w^{m,1}(\Omega)$.

2.1. The Sobolev space $W_w^{m,p}(\Omega)$

We begin by defining the weighted Sobolev space $W_w^{m,p}(\Omega)$. Recall that if $w \in A_p$, then $L_w^p(\Omega) \subset L_{\text{loc}}^1(\Omega)$ for every open set Ω (cf. Remark 1.2.4.1). It thus makes sense to talk about weak derivatives of functions in $L_w^p(\Omega)$.

Definition 2.1.1. Let $\Omega \subset \mathbf{R}^N$ be open, $1 \le p < \infty$, and m a nonnegative integer. Suppose that the weight $w \in A_p$. Then we define the *weighted Sobolev space* $W_w^{m,p}(\Omega)$ as the set of functions $u \in L_w^p(\Omega)$ with weak derivatives $D^\alpha u \in L_w^p(\Omega)$ for $|\alpha| \le m$. The norm of u in $W_w^{m,p}(\Omega)$ is given by

$$\|u\|_{W_w^{m,p}(\Omega)} = \left(\sum_{|\alpha| \le m} \int_\Omega |D^\alpha u|^p w\, dx \right)^{1/p}.$$

We also define $\mathring{W}_w^{m,p}(\Omega)$ as the closure of $C_0^\infty(\Omega)$ in $W_w^{m,p}(\Omega)$.

When $w = 1$, these spaces will be denoted $W^{m,p}(\Omega)$ and $\mathring{W}^{m,p}(\Omega)$, respectively.

The following proposition is proved exactly in the same way as in the non-weighted case (see R. A. Adams [9, pp. 45–46]), using the completeness of $L_w^p(\Omega)$ (see Rudin [95, pp. 69–70]) and the fact that $L_w^p(\Omega) \subset L_{\text{loc}}^1(\Omega)$, when $w \in A_p$.

Proposition 2.1.2. *Let $\Omega \subset \mathbf{R}^N$ be open, $1 \le p < \infty$, and m a nonnegative integer. Suppose that $w \in A_p$. Then $W_w^{m,p}(\Omega)$ and $\mathring{W}_w^{m,p}(\Omega)$ are Banach spaces.*

Another useful consequence of the inclusions $L_w^p(\Omega) \subset L_{\text{loc}}^1(\Omega)$ for general Ω and $L_w^p(\Omega) \subset L^1(\Omega)$ for bounded Ω is the next proposition.

Proposition 2.1.3. *Let $\Omega \subset \mathbf{R}^N$ be open, $1 \le p < \infty$, and m a nonnegative integer. Suppose that $w \in A_p$. Then $W_w^{m,p}(\Omega) \subset W_{\text{loc}}^{m,1}(\Omega)$ and, if Ω is bounded, $W_w^{m,p}(\Omega) \subset W^{m,1}(\Omega)$.*

Here, $W_{\text{loc}}^{m,1}(\Omega)$ denotes the set of functions $u \in L_{\text{loc}}^1(\Omega)$ with weak derivatives $D^\alpha u \in L_{\text{loc}}^1(\Omega)$, $|\alpha| \le m$.

2.1.1. Approximation results

We now turn to the problem of approximating Sobolev functions by smooth functions. The starting point is the theorem below, due to B. Muckenhoupt and R. L. Wheeden [88, pp. 71–74], in the case $p = 1$, and to N. Miller [82, p. 95], in the case $1 < p < \infty$, which shows that functions in $L_w^p(\Omega)$ can be regularized using the standard procedure, if $w \in A_p$. The theorem by Muckenhoupt and Wheeden is, in fact, more general than the one below since they only assume that the function φ belongs to the Schwartz class. We will prove the theorem in the case $p = 1$ under the assumption that φ has compact support, which makes the proof considerably simpler.

Theorem 2.1.4. *Suppose that Ω is an open subset of $\mathbf{R}^N$. Let $\varphi \in C_0^\infty(\mathbf{R}^N)$ be nonnegative such that $\operatorname{supp}\varphi \subset B_1(0)$ and $\int_{\mathbf{R}^N} \varphi\, dx = 1$, and define for $\varepsilon > 0$, $\varphi_\varepsilon(x) = \varepsilon^{-N}\varphi(x/\varepsilon)$, $x \in \mathbf{R}^N$. Suppose that $w \in A_p$, where $1 \le p < \infty$. If $u \in L_w^p(\Omega)$, we define $u_\varepsilon = u * \varphi_\varepsilon$. Then $u_\varepsilon \in C^\infty(\Omega) \cap L_w^p(\Omega)$ with $D^\alpha u_\varepsilon = u * D^\alpha \varphi_\varepsilon$, and as $\varepsilon \to 0$, $u_\varepsilon \to u$ in $L_w^p(\Omega)$.*

Proof. As mentioned above, we will prove the theorem in the case $p = 1$. Since $u \in L_{\text{loc}}^1(\Omega)$, $u_\varepsilon \in C^\infty(\Omega)$. Let $\eta > 0$ be arbitrary and choose a function $\psi \in C_0^1(\Omega)$ such that $\|u - \psi\|_{L_w^1(\Omega)} < \eta$ (see Rudin [95, p. 71]). We then get

$$\|u - u_\varepsilon\|_{L_w^1(\Omega)} < \eta + \|\psi - \psi_\varepsilon\|_{L_w^1(\Omega)} + \|\psi_\varepsilon - u_\varepsilon\|_{L_w^1(\Omega)}. \tag{2.1.1}$$

If $\operatorname{supp}\psi \subset B_R(0)$, then $\operatorname{supp}\psi_\varepsilon \subset B_{2R}(0)$, if $\varepsilon < R$. Also, for every $x \in \mathbf{R}^N$,

$$|\psi(x) - \psi_\varepsilon(x)| \le \int_{|x-y|<\varepsilon} |\psi(x) - \psi(y)|\varphi_\varepsilon(x-y)\, dy \le \sup_{|x-y|<\varepsilon} |\psi(x) - \psi(y)|.$$

Thus, if we choose $\varepsilon < R$ so small that $|\psi(x) - \psi(y)| < \eta/w(B_{2R}(0))$ for every x and y with $|x - y| < \varepsilon$, we see that the second term in (2.1.1) is $< \eta$. Let us now consider the last term in (2.1.1). By Fubini's theorem and Remark 1.2.4.7,

$$\int_{\mathbf{R}^N} |\psi_\varepsilon - u_\varepsilon| w\, dx \le \int_{\mathbf{R}^N} \left(\int_{\mathbf{R}^N} \varphi_\varepsilon(x-y) w(x)\, dx \right) |\psi(y) - u(y)|\, dy$$

$$\le \int_{\mathbf{R}^N} (Mw)|\psi - u|\, dy \le C \int_{\mathbf{R}^N} |\psi - u| w\, dy < C\eta.$$

Here, the second inequality follows from Stein [99, pp. 62–63]. This shows that $u_\varepsilon \to u$, as $\varepsilon \to 0$. Finally, if we replace $\psi - u$ by u in the last series of inequalities, we get $\|u_\varepsilon\|_{L^1_w(\Omega)} \leq C\|u\|_{L^1_w(\Omega)}$ and thus $u_\varepsilon \in L^1_w(\Omega)$. □

Theorem 2.1.4 has the following corollary, which is proved in the same way as when $w = 1$ (see R. A. Adams [9, pp. 52–53]).

Corollary 2.1.5. *Suppose that Ω is an open subset of $\mathbf{R}^N$, and let Ω' be an open subset of Ω such that $\mathrm{dist}(\Omega', \Omega^c) > 0$. Suppose that $u \in W^{m,p}_w(\Omega)$, where $m \geq 1$, $1 \leq p < \infty$, and $w \in A_p$. If we define u_ε as in Theorem 2.1.4, then as $\varepsilon \to 0$, $u_\varepsilon \to u$ in $W^{m,p}_w(\Omega')$. In the case $\Omega = \mathbf{R}^N$, we have convergence in $W^{m,p}_w(\mathbf{R}^N)$.*

Using Corollary 2.1.5, one can prove that $C^\infty(\Omega)$ is dense in $W^{m,p}_w(\Omega)$. In the non-weighted case, this is a result by J. Deny and J. L. Lions [32] for $m = 1$ and by N. G. Meyers and J. Serrin [80] for general m (see R. A. Adams [9, pp. 53–54] or Maz'ya [76, p. 12]).[†]

Corollary 2.1.6. *If $\Omega \subset \mathbf{R}^N$ is open, $m \geq 1$, $1 \leq p < \infty$, and $w \in A_p$, then $C^\infty(\Omega)$ is dense in $W^{m,p}_w(\Omega)$.*

Corollary 2.1.6 implies that functions in $W^{1,p}_w(\Omega)$ can be truncated. The proofs of the proposition and corollary below are identical to the proofs in the case $w = 1$ (see, e.g., Gilbarg–Trudinger [47, pp. 151–152]).

Proposition 2.1.7. *Suppose that $F \in C^1(\mathbf{R})$ with $F' \in L^\infty(\mathbf{R})$. Let $\Omega \subset \mathbf{R}^N$ be open, and let $u \in W^{1,p}_w(\Omega)$, where $1 \leq p < \infty$ and $w \in A_p$. Suppose that $F \circ u \in L^p_w(\Omega)$. Then $F \circ u \in W^{1,p}_w(\Omega)$ with $D_i(F \circ u) = F'(u)D_iu$, $i = 1, \ldots, N$.*

Corollary 2.1.8. *Suppose that $\Omega \subset \mathbf{R}^N$ is open, and let $u \in W^{1,p}_w(\Omega)$, where $1 \leq p < \infty$ and $w \in A_p$. Set*

$$u^+ = \max\{u, 0\}, \quad u^- = \min\{u, 0\}.$$

Then u^+, u^-, and $|u| \in W^{1,p}_w(\Omega)$, and, for $i = 1, \ldots, N$,

$$D_iu^+ = \begin{cases} D_iu, & \text{if } u > 0, \\ 0, & \text{if } u \leq 0, \end{cases}$$

$$D_iu^- = \begin{cases} 0, & \text{if } u \geq 0, \\ D_iu, & \text{if } u < 0, \end{cases}$$

and

$$D_i|u| = \begin{cases} D_iu, & \text{if } u > 0, \\ 0, & \text{if } u = 0, \\ -D_iu, & \text{if } u < 0. \end{cases}$$

[†]This has also been observed by T. Kilpeläinen [67, p. 101].

2.1.2. Extension theorems

The most general type of extension domain for $W_w^{m,p}(\Omega)$ seems to be the class of (ε, δ) domains, defined below. P. W. Jones proved in [62] that if Ω is an (ε, δ) domain, then every function in $W^{m,p}(\Omega)$ can be extended to $W^{m,p}(\mathbf{R}^N)$. Later, S. K. Chua [24] showed that this result remains true for weighted Sobolev spaces if the weight is assumed to be an A_p weight; see also Chua [25], [26].

Definition 2.1.9. Let ε and δ be two positive numbers. An open set $\Omega \subset \mathbf{R}^N$ is called an (ε, δ) *domain*, if, whenever x, $y \in \Omega$ and $|x - y| < \delta$, there is a rectifiable curve $\gamma \subset \Omega$, joining x and y, such that

$$l(\gamma) \leq \frac{1}{\varepsilon}|x - y|$$

and

$$d(z) \geq \varepsilon \frac{|x - z||y - z|}{|x - y|}$$

for all z on γ. Here, $l(\gamma)$ denotes the length of γ and $d(z) = \inf_{a \in \Omega^c} |a - z|$.

Theorem 2.1.10. *Let $\Omega \subset \mathbf{R}^N$ be an (ε, δ) domain, and let m be a positive integer, $1 \leq p < \infty$, and $w \in A_p$. Then there exists a bounded, linear extension operator $\Lambda : W_w^{m,p}(\Omega) \to W_w^{m,p}(\mathbf{R}^N)$. The norm of Λ depends only on ε, δ, m, N, p, and the A_p constant of w.*

In Theorem 2.1.13 below, we show that the classical extension theorem by A. P. Calderón [18] also holds in the weighted situation. Although not as general as Theorem 2.1.10, the proof is much simpler and the theorem suffices for our purposes. Notice, however, that Calderón's construction requires the use of singular integrals and thus only works for $1 < p < \infty$; in the case $p = 1$, we have to rely on Theorem 2.1.10.[†]

We begin by proving two results about singular integrals.

Theorem 2.1.11. *Let $\omega \in C^1(\mathbf{R}^N \setminus \{0\})$ be homogeneous of degree 0 such that*

$$\int_{R_1 < |x| < R_2} \omega(x)\, dx = 0$$

for every R_1 and R_2, $0 < R_1 < R_2 < \infty$. Set $K(x) = |x|^{-N}\omega(x)$, $x \in \mathbf{R}^N \setminus \{0\}$. Suppose that $w \in A_p$, where $1 < p < \infty$. Then, if $f \in L_w^p(\mathbf{R}^N)$, the principal value

$$Tf(x) = \lim_{\varepsilon \to 0} \int_{|x-y| \geq \varepsilon} K(x - y) f(y)\, dy$$

[†]It seems E. M. Stein's construction of extension operators [99, pp. 181–192], which works for domains satisfying a "minimal smoothness condition" and for $1 \leq p \leq \infty$, does not easily generalize to the weighted case.

exists for a.e. $x \in \mathbf{R}^N$, and

$$\|Tf\|_{L^p_w(\mathbf{R}^N)} \leq C\|f\|_{L^p_w(\mathbf{R}^N)}, \tag{2.1.2}$$

with C depending only on $\|\omega\|_{L^\infty(\mathbf{R}^N)}$, $\|\nabla\omega\|_{L^\infty(\mathbf{R}^N)}$, N, p, and the A_p constant of w.

Proof. It is readily seen that the kernel K satisfies

$$|K(x)| \leq \frac{A}{|x|^N}$$

for every $x \in \mathbf{R}^N \setminus \{0\}$, and

$$|K(x) - K(x - y)| \leq \frac{A|y|}{|x|^{N+1}},$$

if x, $y \in \mathbf{R}^N$ and $|x| > 2|y|$ (for the proof of the second inequality, see Stein [99, pp. 39–40]). We define the operator T_ε, $\varepsilon > 0$, and the corresponding maximal operator T^* for $f \in L^p_w(\mathbf{R}^N)$ by

$$T_\varepsilon f(x) = \int_{|x-y|\geq\varepsilon} K(x-y)f(y)\,dy$$

and $T^* f(x) = \sup_{\varepsilon>0} |T_\varepsilon f(x)|$, respectively. Exactly as in the proof of Theorem 1.2.18, one then gets that

$$\|T^*f\|_{L^p_w(\mathbf{R}^N)} \leq C\|f\|_{L^p_w(\mathbf{R}^N)}. \tag{2.1.3}$$

The existence of $Tf(x) = \lim_{\varepsilon\to0} T_\varepsilon f(x)$ a.e. now follows from (2.1.3) using standard techniques (see Stein [99, p. 45]). Finally, the inequality (2.1.2) is a direct consequence of (2.1.3). $\square$

Corollary 2.1.12. *Let the function ω be as in Theorem 2.1.11, let $\delta > 0$, and let $\chi \in C_0^\infty(\mathbf{R}^N)$ with support in $B_{2\delta}(0)$ be such that $\chi = 1$ on $\overline{B_\delta(0)}$. Set $K'(x) = \omega(x)\chi(x)/|x|^N$, $x \in \mathbf{R}^N \setminus \{0\}$. Suppose that $w \in A_p$, where $1 < p < \infty$. Then, if $f \in L^p_w(\mathbf{R}^N)$, the principal value*

$$T'f(x) = \lim_{\varepsilon\to0} \int_{|x-y|\geq\varepsilon} K'(x-y)f(y)\,dy$$

exists for a.e. $x \in \mathbf{R}^N$, and

$$\|T'f\|_{L^p_w(\mathbf{R}^N)} \leq C\|f\|_{L^p_w(\mathbf{R}^N)}, \tag{2.1.4}$$

with C depending only on $\|\omega\|_{L^\infty(\mathbf{R}^N)}$, $\|\nabla\omega\|_{L^\infty(\mathbf{R}^N)}$, $\|\chi\|_{L^\infty(\mathbf{R}^N)}$, N, p, and the A_p constant of w.

Proof. If we let

$$T'_\varepsilon f(x) = \int_{|x-y|\geq\varepsilon} K'(x-y)f(y)\,dy,$$

then, for $0 < \varepsilon < \delta$ and with the notation of the preceding theorem,

$$T'_\varepsilon f(x) = T_\varepsilon f(x) - T_\delta f(x) + \int_{\delta\leq|x-y|<2\delta} K'(x-y)f(y)\,dy,$$

so the existence of $Tf(x)$ a.e. implies the existence of $T'f(x)$. We also have

$$\left|\int_{\delta\leq|x-y|<2\delta} K'(x-y)f(y)\,dy\right| \leq CMf(x).$$

Thus, the inequality (2.1.4) follows from (2.1.2) and Muckenhoupt's theorem (Theorem 1.2.3). $\square$

We now prove the weighted version of Calderón's theorem. The proof follows R. A. Adams [9, pp. 91–94].

Theorem 2.1.13. *Let $w \in A_p$, $1 < p < \infty$, and let $m \geq 1$ be an integer. Suppose that $\Omega \subset \mathbf{R}^N$ is open and satisfies the following cone condition: there is an open covering $\{U_j\}_{j=1}^M$ of $\partial\Omega$ and corresponding finite, congruent cones $\{C_j\}_{j=1}^M$ such that if $x \in U_j \cap \Omega$ for some j, then $x + C_j \subset \Omega$. Then there exists a bounded, linear extension operator $E : W^{m,p}_w(\Omega) \to W^{m,p}_w(\mathbf{R}^N)$. If $u \in W^{m,p}_w(\Omega)$, we thus have*

$$\|Eu\|_{W^{m,p}_w(\mathbf{R}^N)} \leq C\|u\|_{W^{m,p}_w(\Omega)}. \tag{2.1.5}$$

The constant C depends only on Ω, m, N, p, and the A_p constant of w.

Proof. First, let $u \in C^\infty(\Omega) \cap W^{m,p}_w(\Omega)$. Using a partition of unity, it is easy to see that it suffices to construct extension operators $E_j : W^{m,p}_w(\Omega) \to W^{m,p}_w(\mathbf{R}^N)$, $j = 1, \ldots, M$, such that $E_j u = u$ on $U_j \cap \Omega$. Let j be fixed, and set $U = U_j$ and $C = C_j$. Suppose that the length of C is 2δ and that C has its vertex at 0 (for simplicity). Let $\chi \in C_0^\infty(B_{2\delta}(0))$ be such that $\chi \geq 0$ and $\chi = 1$ on $\overline{B_\delta(0)}$, and let $\zeta \in C^\infty(S^{N-1})$ be such that $\zeta \geq 0$ and such that the support of the function φ, $\varphi(x) = \chi(x)\zeta(x/|x|)|x|^{m-N}$, is included in $-C \cup \{0\}$.

We set $D^\alpha u = 0$ in Ω^c and define for $x \in \mathbf{R}^N$,

$$Eu(x) = K\left((-1)^m \int_{S^{N-1}} \left(\int_0^\infty \varphi(\rho\sigma)\rho^{N-1}\frac{\partial^m}{\partial\rho^m}u(x-\rho\sigma)\,d\rho\right)d\sigma\right.$$

$$\left. - \int_{S^{N-1}} \left(\int_0^\infty \psi(\rho\sigma)u(x-\rho\sigma)\,d\rho\right)d\sigma\right)$$

$$= K\left((-1)^m I_1 - I_2\right).$$

Here, K is a constant, whose value will be specified later, the function ψ is given by $\psi(x) = \partial^m(\rho^{N-1}\varphi(x))/\partial\rho^m$, and $x = \rho\sigma$, $0 < \rho < \infty$, $\sigma \in S^{N-1}$, are polar coordinates. Note that $\psi \in C_0^\infty(-C)$ and $\psi = 0$ in $B_\delta(0)$. By integrating I_1 m times by parts with respect to ρ, it follows that $Eu = u$ on $U \cap \Omega$, if K is suitably chosen.

We will now show that E is bounded. We have $I_2 = \theta * u$, where $\theta(x) = \psi(x)|x|^{1-N}$. Since $u \in C^\infty(\Omega)$, we have for $|\alpha| \le m$,

$$|D^\alpha I_2(x)| = |\theta * D^\alpha u(x)| = \left| \int_{\delta \le |x-y| < 2\delta} \frac{\psi(x-y)}{|x-y|^{N-1}} D^\alpha u(y)\, dy \right|$$
$$\le CM(D^\alpha u)(x).$$

Theorem 1.2.3 then implies that $\|D^\alpha I_2\|_{L_w^p(\mathbf{R}^N)} \le C\|D^\alpha u\|_{L_w^p(\Omega)}$. We now consider the integral I_1. According to the chain rule, $I_1 = \sum_{|\alpha|=m} \xi_\alpha * D^\alpha u$, where $\xi_\alpha = (-1)^{|\alpha|} m! \sigma^\alpha \varphi / \alpha!$. We first show that

$$\|D^\beta(\xi_\alpha * v)\|_{L_w^p(\Omega)} \le C\|v\|_{L_w^p(\Omega)} \tag{2.1.6}$$

for $v \in C_0^\infty(\Omega)$ and $|\beta| \le m$. Suppose that $|\beta| \le m - 1$. Then $D^\beta(\xi_\alpha * v) = D^\beta \xi_\alpha * v$, where

$$D^\beta \xi_\alpha(x) = \sum_{k=0}^{|\beta|} \frac{\lambda_k(x) \zeta_k(x/|x|)}{|x|^{N-m+k}},$$

$\lambda_k \in C_0^\infty(B_{2\delta}(0))$, $\lambda_0 = \lambda_1 = ... = \lambda_{|\beta|-1} = 1$ in $B_\delta(0)$, and $\zeta_k \in C_0^\infty(S^{N-1})$. Thus, $D^\beta \xi_\alpha(x) \le CMv(x)$, and (2.1.6) follows as before. If $|\beta| = m$, we write $D^\beta = D^\gamma D_i$, with $|\gamma| = m - 1$. One integration by parts shows that

$$D^\beta(\xi_\alpha * v)(x) = D^\gamma \xi_\alpha * D_i v(x) = \lim_{\varepsilon \to 0} \int_{|x-y| \ge \varepsilon} D^\gamma \xi_\alpha(x-y) D_i v(y)\, dy$$
$$= \lim_{\varepsilon \to 0} \int_{|x-y| \ge \varepsilon} D^\beta \xi_\alpha(x-y) v(y)\, dy + Cv(x).$$

$D^\beta \xi_\alpha$ is homogeneous of degree $-N$ in $B_\delta(0)$ and has mean value 0 on S^{N-1} (since the principal value exists), so (2.1.6) follows from Corollary 2.1.12.

We now have to show that (2.1.6) holds with v replaced by $D^\alpha u$. Let $\{v_n\}$ be a sequence of functions in $C_0^\infty(\Omega)$ such that $v_n \to D^\alpha u$ in $L_w^p(\Omega)$. It then follows from (2.1.6) that there is a subsequence $\{v_{n_i}\}$ such that $D^\beta(\xi_\alpha * v_{n_i}) \to v \in L_w^p(\Omega)$. Now $\xi_\alpha * v_{n_i} \to \xi_\alpha * D^\alpha u$, so we get $v = D^\beta(\xi_\alpha * D^\alpha u)$ and, moreover, that (2.1.6) holds for $u \in W_w^{m,p}(\Omega) \cap C^\infty(\Omega)$.

If $u \in W_w^{m,p}(\Omega)$, we choose a sequence of functions $u_n \in C^\infty(\Omega)$ such that $u_n \to u$ in $W_w^{m,p}(\Omega)$. The inequality (2.1.5), applied to u_n, implies that there is a subsequence $\{u_{n_i}\}$ such that $Eu_{n_i} \to Eu \in W_w^{m,p}(\Omega)$. We immediately see that (2.1.5) holds and, since $u_{n_i} \to u$ and $Eu_{n_i} \to Eu$ a.e., that $Eu = u$ a.e. in Ω, thus completing the proof. $\square$

2.1.3. An interpolation inequality

With the aid of Sobolev's integral formula, we shall now prove an interpolation inequality. A consequence of this result is a Wirtinger type inequality.

Theorem 2.1.14. *Let k and m be positive integers, $0 \le k < m$, and let $w \in A_p$, where $1 \le p < \infty$. Suppose that $B \subset \mathbf{R}^N$ is a ball. Then there is a positive constant C, depending only on k, m, N, p, and the A_p constant of w, such that, for every $u \in W_w^{m,p}(B)$,*

$$\int_B |\nabla^k u|^p w \, dx \le C \left(|B|^{-kp/N} \int_B |u|^p w \, dx + |B|^{(m-k)p/N} \int_B |\nabla^m u|^p w \, dx \right).$$

Proof. It suffices to prove the inequality for $B = B_1(0)$. If $u \in W_w^{m,p}(B)$, then since u, according to Proposition 2.1.3, also belongs to $W^{m,1}(B)$, we may represent u using Sobolev's integral formula (see Maz'ya [76, p. 20] or Adams-Hedberg [7, p. 225]):

$$u(x) = \sum_{|\beta| < m} x^\beta \int_B \varphi_\beta(y) u(y) \, dy + \sum_{|\beta| = m} \int_B \frac{f_\beta(x,y)}{|x-y|^{N-m}} D^\beta u(y) \, dy \quad (2.1.7)$$

for $x \in B$. Here, $\varphi_\beta \in C_0^\infty(B)$, $|\beta| < m$, and $f_\beta \in C^\infty(B \times B)$, $|\beta| = m$, satisfies

$$|D_x^\alpha f_\beta(x,y)| \le \frac{C}{|x-y|^\alpha} \quad (2.1.8)$$

for all multi-indices α. Let α be a multi-index with $|\alpha| = k$. Then,

$$\left| D^\alpha \sum_{|\beta| < m} x^\beta \int_B \varphi_\beta(y) u(y) \, dy \right| \le C \int_B |u| \, dy$$

for $x \in B$. We also have

$$D^\alpha \int_B \frac{f_\beta(x,y)}{|x-y|^{N-m}} D^\beta u(y) \, dy = \int_B D_x^\alpha \left(\frac{f_\beta(x,y)}{|x-y|^{N-m}} \right) D^\beta u(y) \, dy$$

as weak derivatives, so it follows from (2.1.8) and the Leibniz rule that

$$\left| D^\alpha \int_B \frac{f_\beta(x,y)}{|x-y|^{N-m}} D^\beta u(y) \, dy \right| \le C \int_B \frac{|D^\beta u(y)|}{|x-y|^{N-(m-k)}} \, dy.$$

Hence

$$|\nabla^k u(x)| \le C \int_B |u| \, dy + C \int_B \frac{|\nabla^m u(y)|}{|x-y|^{N-(m-k)}} \, dy,$$

and

$$\int_B |\nabla^k u|^p w \, dx \le C w(B) \left(\int_B |u| \, dy \right)^p$$
$$+ C \int_B \left(\int_B \frac{|\nabla^m u(y)|}{|x-y|^{N-(m-k)}} \, dy \right)^p w(x) \, dx.$$

Using Hölder's inequality and the A_p condition (see Remark 1.2.4.1), we obtain

$$\left(\int_B |u| \, dy \right)^p \leq \frac{C}{w(B)} \int_B |u|^p w \, dy.$$

To finish the proof, we have to show that

$$\int_B \left(\int_B \frac{|\nabla^m u(y)|}{|x-y|^{N-(m-k)}} \, dy \right)^p w(x) \, dx \leq C \int_B |\nabla^m u|^p w \, dx. \qquad (2.1.9)$$

First assume that $p = 1$. It is well-known and easy to show that

$$\int_B \frac{w(x)}{|x-y|^{N-(m-k)}} \, dx \leq CMw(y) \qquad (2.1.10)$$

for a.e. $y \in B$ (see Stein [99, pp. 62–63] and Ziemer [108, pp. 85–86] for two different arguments). By Fubini's theorem and the A_1 condition (see Remark 1.2.4.7), we therefore have

$$\int_B \left(\int_B \frac{|\nabla^m u(y)|}{|x-y|^{N-(m-k)}} \, dy \right) w(x) \, dx$$

$$= \int_B \left(\int_B \frac{w(x)}{|x-y|^{N-(m-k)}} \, dx \right) |\nabla^m u(y)| \, dx \leq C \int_B |\nabla^m u| w \, dy.$$

Similarly, for $p > 1$,

$$\int_B \left(\int_B \frac{|\nabla^m u(y)|}{|x-y|^{N-(m-k)}} \, dy \right)^p w(x) \, dx \leq C \int_B (M|\nabla^m u|)^p w \, dx$$

$$\leq C \int_B |\nabla^m u|^p w \, dx, \qquad (2.1.11)$$

where the last inequality follows from Theorem 1.2.3. This proves (2.1.9) and the theorem. □

Theorem 2.1.14 implies the following Wirtinger type inequality. Indeed, if we take P as the polynomial in the right-hand side of (2.1.7), the inequality (2.1.12) follows from the proof of the previous theorem.

Corollary 2.1.15. *Let m be a positive integer and let $w \in A_p$, $1 \leq p < \infty$. Suppose that $B \subset \mathbf{R}^N$ is a ball. Then, if $u \in W_w^{m,p}(B)$, there exists a polynomial $P \in \mathcal{P}_{m-1}$ such that*

$$\sum_{j=0}^{m-1} |B|^{j/N} \|\nabla^j(u-P)\|_{L_w^p(B)} \leq C|B|^{m/N} \|\nabla^m u\|_{L_w^p(B)}, \qquad (2.1.12)$$

where the constant C only depends on m, N, p, and the A_p constant of w.

Remark 2.1.16. The inequality (2.1.12) is a special case of a general Wirtinger inequality proved by S. K. Chua [24, p. 20] by similar methods. Weighted Wirtinger inequalities with first order derivatives in the right-hand side have earlier been studied by E. B. Fabes, C. E. Kenig, and R. P. Serapioni [36, p. 87]. It is easy to see that Corollary 2.1.15 could also be proved using the methods in [36].

2.2. The Sobolev space $V_w^{m,p}(\Omega)$

Recall the following theorem by B. Muckenhoupt and R. L. Wheeden [87, p. 267], mentioned in Section 1.2.7.

Theorem 2.2.1. *Let $0 < \alpha < N$ and $1 < p < N/\alpha$, and set $1/p^* = 1/p - \alpha/N$. Suppose that $w \in A_q$, where $q = p^*/p' + 1$. Then*

$$\left(\int_{\mathbf{R}^N} |\mathcal{I}_\alpha f|^{p^*} w \, dx \right)^{1/p^*} \leq C \left(\int_{\mathbf{R}^N} |f|^p w^{(N-\alpha p)/N} \, dx \right)^{1/p} \tag{2.2.1}$$

for every function $f \in L^p_{w_{\alpha,p}}(\mathbf{R}^N)$, where $w_{\alpha,p} = w^{(N-\alpha p)/N}$. The constant C depends only on α, N, p, and the A_q constant of w. Conversely, if (2.2.1) holds for every $f \in L^p_{w_{\alpha,p}}(\mathbf{R}^N)$, then $w \in A_q$.

An immediate consequence of this theorem, proved by Muckenhoupt and Wheeden [87, pp. 273–274], is the following weighted Sobolev type inequality: let $1/p^* = 1/p - 1/N$, where $1 < p < N$, and suppose that $w \in A_{p^*/p'+1}$. Then, if u is a smooth function on $\mathbf{R}^N$ with compact support,

$$\left(\int_{\mathbf{R}^N} |u|^{p^*} w \, dx \right)^{1/p^*} \leq C \left(\int_{\mathbf{R}^N} |\nabla u|^p w^{(N-p)/N} \, dx \right)^{1/p}. \tag{2.2.2}$$

Indeed, this follows from the representation

$$u(x) = \frac{1}{\omega_{N-1}} \int_{\mathbf{R}^N} \frac{\nabla u(y) \cdot (x-y)}{|x-y|^N} \, dy, \tag{2.2.3}$$

where ω_{N-1} is the $(N-1)$-dimensional measure of the unit sphere S^{N-1} (see Stein [99, p. 125]). More generally, again assuming that u is a smooth function on $\mathbf{R}^N$ with compact support, then since (2.2.3) and M. Riesz' composition formula for Riesz potentials (see Landkof [69, p. 45]) together imply that

$$|u(x)| \leq C \int_{\mathbf{R}^N} \frac{|\nabla^m u(y)|}{|x-y|^{N-m}} \, dy, \tag{2.2.4}$$

it follows from the theorem that if $1 < p < N/m$, then

$$\left(\int_{\mathbf{R}^N} |u|^{p^*} w \, dx \right)^{1/p^*} \leq C \left(\int_{\mathbf{R}^N} |\nabla^m u|^p w^{(N-mp)/N} \, dx \right)^{1/p}, \tag{2.2.5}$$

where the exponent p^* in this case is given by $1/p^* = 1/p - m/N$.

G. David and S. Semmes [30, p. 104] have shown a corresponding result for the case $p = 1$.

Theorem 2.2.2. *Suppose that w is a continuous A_1 weight. Define the exponent 1^* through $1/1^* = 1 - 1/N$. Then there exists a positive constant C such that the inequality*

$$\left(\int_{\mathbf{R}^N} |u|^{1^*} w \, dx \right)^{1/1^*} \leq C \int_{\mathbf{R}^N} |\nabla u| w^{(N-1)/N} \, dx \qquad (2.2.6)$$

holds for every function $u \in C_0^\infty(\mathbf{R}^N)$.

Inspired by these results, we make the following definition.

Definition 2.2.3. Let m be a nonnegative integer, and let $1 \leq p \leq N/m$. Suppose that the weight $w \in A_p$. Then, if Ω is an open subset of $\mathbf{R}^N$, we define the *weighted Sobolev space* $V_w^{m,p}(\Omega)$ as the set of functions $u \in L_w^p(\Omega)$ with weak derivatives $D^\alpha u \in L_{w_{\alpha,p}}^p(\Omega)$, $|\alpha| \leq m$, where $w_{\alpha,p} = w^{(N-|\alpha|p)/N}$. The norm of u in $V_w^{m,p}(\Omega)$ is given by

$$\|u\|_{V_w^{m,p}(\Omega)} = \left(\sum_{|\alpha| \leq m} \int_\Omega |D^\alpha u|^p w_{\alpha,p} \, dx \right)^{1/p}.$$

We also define $\overset{\circ}{V}_w^{m,p}(\Omega)$ as the closure of $C_0^\infty(\Omega)$ in $V_w^{m,p}(\Omega)$.

We will mainly concentrate on the case $p > 1$ in this section. Using a Sobolev type inequality for balls (Proposition 2.2.9), we show in Theorem 2.2.10 that (2.2.5) holds for functions belonging to $V_w^{m,p}(\mathbf{R}^N)$. We will also show in Theorem 2.2.20, perhaps a bit surprisingly, that $V_w^{m,p}(\Omega) = W_{w_{m,p}}^{m,p}(\Omega)$, if Ω is a bounded extension domain for both spaces, e.g., if Ω satisfies a cone condition. This result together with (2.2.5) indicates that the space $V_w^{m,p}(\Omega)$ may be of use in the study of degenerate elliptic partial differential equations on unbounded domains if the degeneration is of the form $w^{(N-mp)/N}$, where $w \in A_p$. Finally, we prove that truncation acts on $V_w^{1,p}(\Omega)$. This fact will be used in Section 2.4 for the proof of a two-weighted isoperimetric inequality. In Section 2.5, we extend (2.2.6) to the case of non-smooth weights and show how this implies Sobolev type inequalities for $\overset{\circ}{W}_w^{m,1}(\Omega)$, where Ω is bounded, and $W_w^{m,1}(\Omega)$, where Ω is a bounded (ε, δ) domain.

We begin by proving a lemma, a consequence of which is that $w^s \in A_p$ for $0 \leq s \leq 1$, if $w \in A_p$. This is quite natural in view of the fact that the weight w^s has "weaker" zeros and singularities than w, both locally and globally. In particular, the weights $w_{\alpha,p}$, occurring in the norm of $V_w^{m,p}(\Omega)$, are all A_p weights.

Lemma 2.2.4. *If $w \in A_p$, $1 \leq p < \infty$, with A_p constant C, and $0 \leq s \leq 1$, then $w^s \in A_q$, where $q = s(p-1) + 1$, with A_q constant $\leq C^s$.*

Proof. Let B be an arbitrary ball. If $1 < p < \infty$, then Hölder's inequality and the A_p condition imply that

$$\left(\fint_B w^s \, dx \right) \left(\fint_B (w^s)^{-1/(q-1)} \, dx \right)^{q-1}$$
$$\leq \left(\left(\fint_B w \, dx \right) \left(\fint_B w^{-1/(p-1)} \, dx \right)^{p-1} \right)^s \leq C^s,$$

and if $p = 1$, then

$$\left(\fint_B w^s \, dx \right) \operatorname*{ess\,sup}_{x \in B} \frac{1}{w(x)^s} \leq \left(\left(\fint_B w \, dx \right) \operatorname*{ess\,sup}_{x \in B} \frac{1}{w(x)} \right)^s \leq C^s. \quad \square$$

The space $V_w^{m,p}(\Omega)$ shares the following two basic properties with $W_w^{m,p}(\Omega)$ (cf. Proposition 2.1.2 and Proposition 2.1.3).

Proposition 2.2.5. *Let Ω be an open subset of $\mathbf{R}^N$, m a nonnegative integer, and $1 \leq p \leq N/m$. Suppose that $w \in A_p$. Then $V_w^{m,p}(\Omega)$ and $\overset{\circ}{V}_w^{m,p}(\Omega)$ are Banach spaces.*

Proposition 2.2.6. *Let Ω be an open subset of $\mathbf{R}^N$, m a nonnegative integer, and $1 \leq p \leq N/m$. Suppose that $w \in A_p$. Then $V_w^{m,p}(\Omega) \subset W_{\text{loc}}^{m,1}(\Omega)$ and, if Ω is bounded, $V_w^{m,p}(\Omega) \subset W^{m,1}(\Omega)$.*

The corollary below follows in the same way as the inequality (2.2.5) from Theorem 2.2.1.

Corollary 2.2.7. *Let $m \geq 1$ be an integer, and let $1 < p < N/m$. Let k be an integer, $0 \leq k < m$, and set $1/p^* = 1/p - (m-k)/N$. Suppose that $w \in A_p$. Then there exists a positive constant C, depending only on k, m, N, p, and the A_p constant of w, such that, for every $u \in \overset{\circ}{V}_w^{m,p}(\mathbf{R}^N)$,*

$$\left(\int_{\mathbf{R}^N} |\nabla^k u|^{p^*} w_{k,p^*} \, dx \right)^{1/p^*} \leq C \left(\int_{\mathbf{R}^N} |\nabla^m u|^p w_{m,p} \, dx \right)^{1/p}. \tag{2.2.7}$$

Proof. Note that, by Lemma 2.2.4, the weight

$$w_{k,p^*} = w^{(N-kp^*)/N} = w^{(N-mp)/(N-(m-k)p)}$$

belongs to A_p. We assume, as we may, that $u \in C_0^\infty(\mathbf{R}^N)$. If we then use the fact that

$$|\nabla^k u(x)| \leq C \mathcal{I}_{m-k} |\nabla^m u(x)|, \tag{2.2.8}$$

which follows from (2.2.4), and Theorem 2.2.1, we obtain (2.2.7):

$$\left(\int_{\mathbf{R}^N} |\nabla^k u|^{p^*} w_{k,p^*}\, dx\right)^{1/p^*} \leq \left(\int_{\mathbf{R}^N} (\mathcal{I}_{m-k}|\nabla^m u|)^{p^*} w_{k,p^*}\, dx\right)^{1/p^*}$$

$$\leq C \left(\int_{\mathbf{R}^N} |\nabla^m u|^p w_{k,p^*}^{(N-(m-k)p)/N}\, dx\right)^{1/p}$$

$$= C \left(\int_{\mathbf{R}^N} |\nabla^m u|^p w_{m,p}\, dx\right)^{1/p}. \quad \square$$

We next prove a limiting case version of Theorem 2.2.1.

Lemma 2.2.8. *Let* $0 < \alpha < N$ *and* $1 < p < \infty$, *and suppose that* $\alpha p = N$. *Suppose that* $w \in A_\infty$, *and let* $1 \leq q < \infty$. *Then there exists a positive constant* C, *depending only on* α, N, p, q, *and the* A_∞ *constants of* w, *such that if* Ω *is a bounded, open subset of* $\mathbf{R}^N$ *and* $f \in L^p(\Omega)$, *then*

$$\left(\int_\Omega |\mathcal{I}_\alpha f|^q w\, dx\right)^{1/q} \leq C w(\Omega)^{1/q} \left(\int_\Omega |f|^p\, dx\right)^{1/p}. \tag{2.2.9}$$

Proof. By Theorem 1.2.10, we know that $w \in A_s$ for some s, $1 < s < \infty$. First, suppose that $q \geq q_0 = sp'$ and define r by $1/r = 1/q + \alpha/N$. Then $\alpha r < N$, $q = r^*$, and $r^*/r' + 1 \geq s$, so $w \in A_{r^*/r'+1}$. Theorem 2.2.1 now implies that

$$\left(\int_\Omega |\mathcal{I}_\alpha f|^q w\, dx\right)^{1/q} = \left(\int_\Omega |\mathcal{I}_\alpha f|^{r^*} w\, dx\right)^{1/r^*}$$

$$\leq C \left(\int_\Omega |f|^r w_{\alpha,r}\, dx\right)^{1/r}. \tag{2.2.10}$$

If we then apply Hölder's inequality, with exponent $N/\alpha r = p/r$, to the last integral, we get

$$\left(\int_\Omega |f|^r w_{\alpha,r}\, dx\right)^{1/r} \leq C w(\Omega)^{(N-\alpha r)/Nr} \left(\int_\Omega |f|^p\, dx\right)^{1/p}$$

$$= C w(\Omega)^{1/q} \left(\int_\Omega |f|^p\, dx\right)^{1/p},$$

and this, together with (2.2.10), gives the inequality (2.2.9).

The case $1 \leq q < q_0$ follows from Hölder's inequality using (2.2.9) with exponent q_0. $\square$

The next proposition is a Sobolev type inequality for $V_w^{m,p}(B)$, where B is a ball, which we will use several times in the rest of this section. The first application is in the proof of Theorem 2.2.10, from which it follows that the inequality (2.2.7) holds for functions in $V_w^{m,p}(\mathbf{R}^N)$.

Proposition 2.2.9. *Let $m \geq 1$ be an integer and let $1 < p \leq N/m$. Suppose that $w \in A_p$. Let k be an integer, $0 \leq k < m$, and set $1/p^* = 1/p - (m-k)/N$. Let q be such that $1 \leq q \leq p^*$, when $(m-k)p < N$, and $1 \leq q < \infty$, when $(m-k)p = N$. Then there exists a positive constant C, depending only on k, m, N, p, q, and the A_p constant of w, such that if $B \subset \mathbf{R}^N$ is a ball and $u \in V_w^{m,p}(B)$, then*

$$\left(\int_B |\nabla^k u|^q w_{k,q} \, dx \right)^{1/q} \leq C w(B)^{1/q - 1/p^*} \left(w(B)^{-m/N} \left(\int_B |u|^p w \, dx \right)^{1/p} \right.$$
$$\left. + \left(\int_B |\nabla^m u|^p w_{m,p} \, dx \right)^{1/p} \right). \tag{2.2.11}$$

Proof. The proof is similar to that of Theorem 2.1.14. We may assume that $B = B_1(0)$. If α is a multi-index with $|\alpha| = k$, then, by Sobolev's formula,

$$D^\alpha u(x) = \sum_{|\beta| < m} D^\alpha x^\beta \int_B \varphi_\beta(y) u(y) \, dy$$
$$+ \sum_{|\beta| = m} \int_B D_x^\alpha \left(\frac{f_\beta(x,y)}{|x-y|^{N-m}} \right) D^\beta u(y) \, dy.$$

It follows that

$$\left(\int_B |\nabla^k u|^q w_{k,q} \, dx \right)^{1/q} \leq C \left(\int_B w_{k,q} \, dx \right)^{1/q} \int_B |u| \, dy$$
$$+ C \left(\int_B (\mathcal{I}_{m-k} |\nabla^m u|)^q w_{k,q} \, dx \right)^{1/q}. \tag{2.2.12}$$

The first term in the right-hand side of (2.2.12) is estimated using Hölder's inequality and the A_p condition as follows:

$$\left(\int_B w_{k,q} \, dx \right)^{1/q} \int_B |u| \, dy \leq C w(B)^{1/q - k/N} w(B)^{-1/p} \left(\int_B |u|^p w \, dx \right)^{1/p}$$
$$= C w(B)^{1/q - 1/p^*} w(B)^{-m/N} \left(\int_B |u|^p w \, dx \right)^{1/p}.$$

To handle the second term in the right-hand side of (2.2.12), we first assume that $(m-k)p < N$. We split $w_{k,q}$ into the product

$$w^{(N-kp^*)q/Np^*} w^{(N-kq)/N - (N-kp^*)q/Np^*} = w_{k,p^*}^{q/p^*} w^{1/(p^*/q)'}$$

and apply Hölder's inequality with exponent p^*/q, to obtain

$$\left(\int_B (\mathcal{I}_{m-k} |\nabla^m u|)^q w_{k,q} \, dx \right)^{1/q}$$

$$\leq w(B)^{1/q-1/p^*} \left(\int_B (\mathcal{I}_{m-k}|\nabla^m u|)^{p^*} w_{k,p^*}\, dx \right)^{1/p^*}$$

$$\leq Cw(B)^{1/q-1/p^*} \left(\int_B |\nabla^m u|^p w_{m,p}\, dx \right)^{1/p}, \qquad (2.2.13)$$

where the last inequality follows from Theorem 2.2.1. If $(m-k)p = N$, then $k = 0$ and $mp = N$ since $mp \leq N$, so by Lemma 2.2.8,

$$\left(\int_B (\mathcal{I}_m|\nabla^m u|)^q w\, dx \right)^{1/q} \leq Cw(B)^{1/q} \left(\int_B |\nabla^m u|^p\, dx \right)^{1/p}. \quad \square$$

Theorem 2.2.10. *Let $m \geq 1$ be an integer and let $1 < p \leq N/m$. Suppose that $w \in A_p$. Let k be an integer, $0 \leq k < m$, and set $1/p^* = 1/p - (m-k)/N$. Let q be such that $p \leq q \leq p^*$, when $(m-k)p < N$, and $p \leq q < \infty$, when $(m-k)p = N$. Then there exists a positive constant C, depending only on k, m, N, p, q, and the A_p constant of w, such that if $u \in V_w^{m,p}(\mathbf{R}^N)$, then*

$$\|\nabla^k u\|_{L^q_{w_{k,q}}(\mathbf{R}^N)} \leq C\|u\|_{L^p_w(\mathbf{R}^N)}^{N(1/q-1/p^*)/m} \|\nabla^m u\|_{L^p_{w_{m,p}}(\mathbf{R}^N)}^{1-N(1/q-1/p^*)/m}. \qquad (2.2.14)$$

Remark 2.2.11. *Thus, if $mp < N$ and $q = p^*$, then*

$$\left(\int_{\mathbf{R}^N} |\nabla^k u|^{p^*} w_{k,p^*}\, dx \right)^{1/p^*} \leq C \left(\int_{\mathbf{R}^N} |\nabla^m u|^p w_{m,p}\, dx \right)^{1/p}.$$

Note also that, by Young's inequality, the right-hand side of (2.2.14) may be replaced by a constant times

$$\|u\|_{L^p_w(\mathbf{R}^N)} + \|\nabla^m u\|_{L^p_{w_{m,p}}(\mathbf{R}^N)}.$$

The proof below is a minor modification of the proof given in the book by V. G. Maz'ya [76, p. 67].

Proof. Let $R > 0$ be arbitrary, but fixed, and let $x \in B_R(0) \cap \operatorname{supp}|\nabla^k u|$. By Proposition 2.2.9, we know that

$$\|\nabla^k u\|_{L^q_{w_{k,q}}(B_\rho(x))} \leq Cw(B_\rho(x))^\alpha \|u\|_{L^p_w(B_\rho(x))}$$

$$+ Cw(B_\rho(x))^\beta \|\nabla^m u\|_{L^p_{w_{m,p}}(B_\rho(x))}, \qquad (2.2.15)$$

for arbitrary $\rho > 0$, where $\alpha = 1/q - 1/p^* - m/N \leq 0$ and $\beta = 1/q - 1/p^* \geq 0$, and not both α and β equal 0. Let ρ_0, $0 < \rho_0 < R$, be arbitrary. We consider two cases. First, if

$$w(B_\rho(x))^\alpha \|u\|_{L^p_w(B_\rho(x))} \leq w(B_\rho(x))^\beta \|\nabla^m u\|_{L^p_{w_{m,p}}(B_\rho(x))} \qquad (2.2.16)$$

for $\rho = \rho_0$, we fix the value of ρ to ρ_0. Then

$$w(B_\rho(x)) \geq \left(\frac{\|u\|_{L_w^p(B_\rho(x))}}{\|\nabla^m u\|_{L_{w_{m,p}}^p(B_\rho(x))}} \right)^{N/m}$$

(we note that the norm in the denominator is positive by the assumption that $x \in B_R(0) \cap \operatorname{supp} |\nabla^k u|$ together with (2.2.15) and (2.2.16)), so by (2.2.15),

$$\|\nabla^k u\|_{L_{w_{k,q}}^q(B_\rho(x))} \leq C \|u\|_{L_w^p(B_\rho(x))}^{N\beta/m} \|\nabla^m u\|_{L_{w_{m,p}}^p(B_\rho(x))}^{1-N\beta/m}$$
$$+ C w(B_{\rho_0}(x))^\beta \|\nabla^m u\|_{L_{w_{m,p}}^p(B_\rho(x))}.$$

On the other hand, if (2.2.16) does not hold, we use the fact that $\int_{\mathbf{R}^N} w \, dx = \infty$ (see Remark 1.2.11) and increase ρ until the left-hand side equals the right-hand side. In this case, (2.2.15) implies that

$$\|\nabla^k u\|_{L_{w_{k,q}}^q(B_\rho(x))} \leq C \|u\|_{L_w^p(B_\rho(x))}^{N\beta/m} \|\nabla^m u\|_{L_{w_{m,p}}^p(B_\rho(x))}^{1-N\beta/m}$$
$$\leq C \|u\|_{L_w^p(B_\rho(x))}^{N\beta/m} \|\nabla^m u\|_{L_{w_{m,p}}^p(B_\rho(x))}^{1-N\beta/m}$$
$$+ C w(B_{\rho_0}(x))^\beta \|\nabla^m u\|_{L_{w_{m,p}}^p(B_\rho(x))}.$$

Recall that, by Theorem 1.2.9, $w \in A_\infty$, so there are positive constants C and δ such that

$$w(B_{\rho_0}(x)) \leq C \left(\frac{|B_{\rho_0}(x)|}{|B_{2R}(0)|} \right)^\delta w(B_{2R}(0)) = C C_R \rho_0^{N\delta},$$

where $C_R = R^{-N\delta} w(B_{2R}(0))$. Thus, for every $x \in B_R(0) \cap \operatorname{supp} |\nabla^k u|$,

$$\|\nabla^k u\|_{L_{w_{k,q}}^q(B_\rho(x))} \leq C \|u\|_{L_w^p(B_\rho(x))}^{N\beta/m} \|\nabla^m u\|_{L_{w_{m,p}}^p(B_\rho(x))}^{1-N\beta/m}$$
$$+ C C_R^\beta \rho_0^{\beta\delta N} \|\nabla^m u\|_{L_{w_{m,p}}^p(B_\rho(x))} \tag{2.2.17}$$

for some number $\rho = \rho(x)$. The Besicovitch theorem (see, e.g., Maz'ya [76, pp. 30–32]) now provides us with balls B_j, satisfying (2.2.17), such that $\{B_j\}$ covers $B_R(0) \cap \operatorname{supp} |\nabla^k u|$ with finite multiplicity, the multiplicity depending only on N. Using this and Hölder's inequality, we obtain

$$\|\nabla^k u\|_{L_{w_{k,q}}^q(B_R(0))}^p \leq C \sum_j (\|u\|_{L_w^p(B_j)}^p)^{N\beta/m} (\|\nabla^m u\|_{L_{w_{m,p}}^p(B_j)}^p)^{1-N\beta/m}$$
$$+ C C_R^{\beta p} \rho_0^{\beta\delta Np} \sum_j \|\nabla^m u\|_{L_{w_{m,p}}^p(B_j)}^p$$
$$\leq C (\|u\|_{L_w^p(\mathbf{R}^N)}^{N\beta/m} \|\nabla^m u\|_{L_{w_{m,p}}^p(\mathbf{R}^N)}^{1-N\beta/m})^p$$
$$+ C C_R^{\beta p} \rho_0^{\beta\delta Np} \|\nabla^m u\|_{L_{w_{m,p}}^p(\mathbf{R}^N)}^p.$$

If we finally let $\rho_0 \to 0$ and $R \to \infty$ (in that order) in the last inequality, we arrive at (2.2.14). $\square$

We will next prove that the space $V_w^{m,p}(\Omega)$ has the usual approximation properties. The first result (Proposition 2.2.12), concerning local approximation with smooth functions, is proved in the same way as in the non-weighted case. For the proof of the second result (Proposition 2.2.13), which shows that $C^\infty(\Omega)$ is dense in $V_w^{m,p}(\Omega)$, we need Lemma 2.2.14. This lemma enables us to control the lower order derivatives that occur when the Leibniz rule is applied to products (cf. the proof of Proposition 2.2.13). The lemma also implies that $V_w^{m,p}(B) \subset W_{w_{m,p}}^{m,p}(B)$ for balls B and will be used later in Theorem 2.2.20 to show that the same inclusion holds for bounded extension domains.

Proposition 2.2.12. *Suppose that Ω is an open subset of $\mathbf{R}^N$, and let Ω' be an open subset of Ω such that $\mathrm{dist}(\Omega', \Omega^c) > 0$. Suppose that $u \in V_w^{m,p}(\Omega)$, where $m \geq 1$, $1 \leq p \leq N/m$, and $w \in A_p$. If we define u_ε as in Theorem 2.1.4, then $u_\varepsilon \to u$ in $V_w^{m,p}(\Omega')$, as $\varepsilon \to 0$. In the case $\Omega = \mathbf{R}^N$, we have convergence in $V_w^{m,p}(\mathbf{R}^N)$.*

Proposition 2.2.13. *Let $m \geq 1$ be an integer, and let $1 \leq p \leq N/m$. Suppose that $w \in A_p$. Then, if Ω is an open subset of $\mathbf{R}^N$, $C^\infty(\Omega) \cap V_w^{m,p}(\Omega)$ is dense in $V_w^{m,p}(\Omega)$.*

Lemma 2.2.14. *Let $m \geq 1$ be an integer, and let $1 < p \leq N/m$. Suppose that $w \in A_p$. Let k be an integer, $0 \leq k < m$. Then there exists a positive constant C, depending only on k, m, N, p, and the A_p constant of w, such that if $B \subset \mathbf{R}^N$ is a ball and $u \in V_w^{m,p}(B)$, then*

$$\left(\int_B |\nabla^k u|^p w_{m,p} \, dx \right)^{1/p} \leq C|B|^{(m-k)/N} \left(w(B)^{-m/N} \left(\int_B |u|^p w \, dx \right)^{1/p} \right.$$
$$\left. + \left(\int_B |\nabla^m u|^p w_{m,p} \, dx \right)^{1/p} \right).$$

Proof. In the case $(m-k)p < N$, we note that $w_{m,p} = w_{k,p^*}^{p/p^*}$, where p^* in this case is defined by $1/p^* = 1/p - (m-k)/N$, and then use Hölder's inequality with exponent p^*/p and Proposition 2.2.9, to obtain

$$\left(\int_B |\nabla^k u|^p w_{m,p} \, dx \right)^{1/p} \leq |B|^{(m-k)/N} \left(\int_B |\nabla^k u|^{p^*} w_{k,p^*} \, dx \right)^{1/p^*}$$
$$\leq C|B|^{(m-k)/N} \left(w(B)^{-m/N} \left(\int_B |u|^p w \, dx \right)^{1/p} \right.$$
$$\left. + \left(\int_B |\nabla^m u|^p w_{m,p} \, dx \right)^{1/p} \right).$$

If $(m-k)p = N$, we have $mp = N$ and $k = 0$. By writing $1 = w^{1/p}w^{-1/p}$ and applying Hölder's inequality with exponent p, it follows from the A_p condition and Proposition 2.2.9 that

$$\left(\int_B |u|^p\, dx\right)^{1/p} \leq \left(\int_B |u|^{p^2} w\, dx\right)^{1/p^2} \left(\int_B w^{-1/(p-1)}\, dx\right)^{(p-1)/p^2}$$

$$\leq C\left(\frac{|B|^p}{w(B)}\right)^{1/p^2} w(B)^{1/p^2}\left(w(B)^{-m/N}\left(\int_B |u|^p w\, dx\right)^{1/p}\right.$$

$$\left. + \left(\int_B |\nabla^m u|^p w_{m,p}\, dx\right)^{1/p}\right)$$

$$= C|B|^{m/N}\left(w(B)^{-m/N}\left(\int_B |u|^p w\, dx\right)^{1/p}\right.$$

$$\left. + \left(\int_B |\nabla^m u|^p w_{m,p}\, dx\right)^{1/p}\right),$$

thus proving the lemma. $\square$

Proof of Proposition 2.2.13. Our proof follows Maz'ya [76, p. 12]. We let $\{B_j\}_{j=1}^\infty$ be a locally finite covering of Ω with balls B_j, satisfying $\bar{B}_j \subset \Omega$, and then let $\{\psi_j\}_{j=1}^\infty$ be a partition of unity, subordinate to $\{B_j\}_{j=1}^\infty$. We first claim that $\psi_j u \in V_w^{m,p}(\Omega)$ with support in B_j. By the Leibniz rule, it suffices to prove that the integrals

$$\int_{B_j} |D^\alpha u|^p w_{k,p}\, dx$$

are finite for $|\alpha| \leq k$ and $0 \leq k \leq m$. In the case $p > 1$, this follows from Lemma 2.2.14, and when $p = 1$, we have

$$\int_{B_j} |D^\alpha u| w_{k,1}\, dx = \int_{B_j} |D^\alpha u| w^{(N-|\alpha|)/N} w^{(|\alpha|-k)/N}\, dx$$

$$\leq C\left(\frac{|B_j|}{w(B_j)}\right)^{(k-|\alpha|)/N} \int_{B_j} |D^\alpha u| w_{\alpha,1}\, dx,$$

since $w \in A_1$, and the last integral is finite by definition. This proves the claim.

Now let $\eta > 0$ be arbitrary, and let φ_ε be the mollifier in Theorem 2.1.4. Define v_j by $v_j = \varphi_{\varepsilon_j} * (\psi_j u)$, $j = 1, 2, \ldots$, where the $\varepsilon_j > 0$ are chosen so that $\operatorname{supp} v_j \subset B_j$ and

$$\|\psi_j u - v_j\|_{V_w^{m,p}(B_j)} < 2^{-j}\eta.$$

If we then set $v = \sum_{j=1}^\infty v_j$, it follows from the local finiteness of the covering $\{B_j\}_{j=1}^\infty$ that $v \in C^\infty(\Omega)$. Furthermore,

$$\|u - v\|_{V_w^{m,p}(\Omega)} \leq \sum_{j=1}^\infty \|\psi_j u - v_j\|_{V_w^{m,p}(B_j)} < \eta. \quad \square$$

Proposition 2.2.13 implies the following counterparts to Corollary 2.1.8 and Theorem 2.1.13.

Corollary 2.2.15. *Suppose that $\Omega \subset \mathbf{R}^N$ is open, and let $u \in V_w^{1,p}(\Omega)$, where $1 \leq p \leq N$ and $w \in A_p$. Set*

$$u^+ = \max\{u, 0\}, \quad u^- = \min\{u, 0\}.$$

Then u^+, u^-, and $|u| \in V_w^{1,p}(\Omega)$, and, for $i = 1, \ldots, N$,

$$D_i u^+ = \begin{cases} D_i u, & \text{if } u > 0, \\ 0, & \text{if } u \leq 0, \end{cases}$$

$$D_i u^- = \begin{cases} 0, & \text{if } u \geq 0, \\ D_i u, & \text{if } u < 0, \end{cases}$$

and

$$D_i |u| = \begin{cases} D_i u, & \text{if } u > 0, \\ 0, & \text{if } u = 0, \\ -D_i u, & \text{if } u < 0. \end{cases}$$

Theorem 2.2.16. *Let m be a positive integer and $1 < p \leq N/m$, and suppose that $w \in A_p$. Let Ω be an open subset of $\mathbf{R}^N$ which satisfies the following cone condition: there is a open covering $\{U_j\}_{j=1}^M$ of $\partial\Omega$ and corresponding finite, congruent cones $\{C_j\}_{j=1}^M$ such that if $x \in U_j \cap \Omega$, then $x + C_j \subset \Omega$. Then there exists a bounded, linear extension operator $E : V_w^{m,p}(\Omega) \to V_w^{m,p}(\mathbf{R}^N)$. If $u \in V_w^{m,p}(\Omega)$, we thus have*

$$\|Eu\|_{V_w^{m,p}(\mathbf{R}^N)} \leq C \|u\|_{V_w^{m,p}(\Omega)}.$$

The constant C depends only on Ω, m, N, p, and the A_p constant of w.

Let us end this section by proving that $V_w^{m,p}(\Omega) = W_{w_{m,p}}^{m,p}(\Omega)$ for bounded extension domains Ω, when $1 < p \leq N/m$. For the proof, we need some lemmas, the first of which, a reverse Hölder inequality, is due to J.-O. Strömberg and R. L. Wheeden [101, p. 347].

Lemma 2.2.17. *If $w \in A_p$, where $1 \leq p < \infty$, and $0 < s < 1$, then there exists a positive constant C, depending only on p and s, such that*

$$\fint_B w \, dx \leq \frac{A}{C} \left(\fint_B w^s \, dx \right)^{1/s} \tag{2.2.18}$$

for every ball $B \subset \mathbf{R}^N$, where A is the A_p constant of w.

In the proof of Lemma 2.2.17, we use a result due to R. Coifman and C. Fefferman ([28, p. 247]) which follows easily from the A_p condition.

Lemma 2.2.18. *Let $1 < p < \infty$. If $w \in A_p$, with A_p constant A, and the number β satisfies $0 < \beta \leq 1/A$, then*

$$\left|\left\{ x \in B \,;\, w(x) \geq \beta \fint_B w\, dy \right\}\right| \geq (1 - (A\beta)^{1/(p-1)})|B| \qquad (2.2.19)$$

for every ball $B \subset \mathbf{R}^N$.

Proof of Lemma 2.2.17. The proof of (2.2.18) is particularly simple when $p = 1$ and $w \in A_1$, since then

$$w(x) \geq \frac{1}{A} \fint_B w\, dy$$

for a.e. $x \in B$. Now suppose that $1 < p < \infty$. If E is the set in the left-hand side of (2.2.19) with $\beta = 1/2A$, then

$$\left(\fint_B w^s\, dx\right)^{1/s} \geq \frac{|E|}{|B|^{1/s}}\left(\fint_E w^s\, dx\right)^{1/s} \geq \left(\frac{|E|}{|B|}\right)^{1/s}\frac{1}{2A}\fint_B w\, dy$$

$$\geq \frac{1}{2A}(1 - 2^{-1/(p-1)})^{1/s}\fint_B w\, dy. \quad \square$$

Lemma 2.2.14 implies that $W_{w_{m,p}}^{m,p}(B) \subset V_w^{m,p}(B)$, when B is a ball. The next lemma shows that the reverse inclusion also holds, so that $W_{w_{m,p}}^{m,p}(B) = V_w^{m,p}(B)$.

Lemma 2.2.19. *Let $m \geq 1$ be an integer, and let $1 < p \leq N/m$. Suppose that $w \in A_p$. Let k be an integer, $0 \leq k < m$. Then there exists a positive constant C, depending only on k, m, N, p, and the A_p constant of w, such that if $B \subset \mathbf{R}^N$ is a ball and $u \in W_{w_{m,p}}^{m,p}(B)$, then*

$$\left(\int_B |\nabla^k u|^p w_{k,p}\, dx\right)^{1/p} \leq C w(B)^{(m-k)/N}\left(|B|^{-m/N}\left(\int_B |u|^p w_{m,p}\, dx\right)^{1/p}\right.$$

$$\left. + \left(\int_B |\nabla^m u|^p w_{m,p}\, dx\right)^{1/p}\right).$$

Proof. The proof is a variation of the proof of Proposition 2.2.9. We may assume that $B = B_1(0)$. By Sobolev's formula,

$$\left(\int_B |\nabla^k u|^p w_{k,p}\, dx\right)^{1/p} \leq C\left(\int_B w_{k,p}\, dx\right)^{1/p}\int_B |u|\, dy$$

$$+ C\left(\int_B (\mathcal{I}_{m-k}|\nabla^m u|)^p\, w_{k,p}\, dx\right)^{1/p}.$$

To bound the first term in the right-hand side, we use Hölder's inequality together with the A_p condition (cf. Remark 1.2.4.1) and Lemma 2.2.17:

$$\left(\int_B w_{k,p}\, dx \right)^{1/p} \int_B |u|\, dy$$

$$\leq Cw(B)^{(N-kp)/Np} \left(\int_B w_{m,p}\, dx \right)^{-1/p} \left(\int_B |u|^p w_{m,p}\, dx \right)^{1/p}$$

$$\leq Cw(B)^{(m-k)/N} \left(\int_B |u|^p w_{m,p}\, dx \right)^{1/p}$$

(note that this is true also when $mp = N$). For the second term, we first assume that $(m-k)p < N$. The inequality (2.2.13) above with $q = p$ then gives

$$\left(\int_B (\mathcal{I}_{m-k}|\nabla^m u|)^p\, w_{k,p}\, dx \right)^{1/p} \leq Cw(B)^{(m-k)/N} \left(\int_B |\nabla^m u|^p w_{m,p}\, dx \right)^{1/p}.$$

In the case $(m-k)p = N$, the desired inequality instead follows directly from the inequality (2.2.13). $\square$

We are now ready to prove the equality between $V_w^{m,p}(\Omega)$ and $W_{w_{m,p}}^{m,p}(\Omega)$. We formulate the result for domains satisfying the cone condition in Theorem 2.2.16 since we know that such domains are extension domains for both spaces.

Theorem 2.2.20. *Let $m \geq 1$ be an integer, and let $1 < p \leq N/m$. Suppose that $w \in A_p$. Let Ω be an open, bounded subset of $\mathbf{R}^N$, which satisfies the cone condition in Theorem 2.2.16. Then $V_w^{m,p}(\Omega) = W_{w_{m,p}}^{m,p}(\Omega)$.*

Proof. We will show that $V_w^{m,p}(\Omega) \subset W_{w_{m,p}}^{m,p}(\Omega)$; the opposite inclusion is proved similarly. Suppose that $\Omega \subset B_R(a)$, and let $\psi \in C_0^\infty(\mathbf{R}^N)$ with $\psi = 1$ on $B_1(0)$ and $\operatorname{supp} \psi \subset B_2(0)$. Set $\varphi(x) = \psi((x-a)/R)$, $x \in \mathbf{R}^N$. If $u \in V_w^{m,p}(\Omega)$, we let Eu be a bounded extension of u to $V_w^{m,p}(\mathbf{R}^N)$ and set $v = \varphi Eu$. By Lemma 2.2.19, we then have

$$\int_\Omega |\nabla^k u|^p w_{m,p}\, dx \leq \int_{B_{2R}(a)} |\nabla^k v|^p w_{m,p}\, dx$$

$$\leq C|B_{2R}(a)|^{(m-k)p/N} \left(w(B_{2R}(a))^{-mp/N} \int_{B_{2R}(a)} |v|^p w\, dx \right.$$

$$\left. + \int_{B_{2R}(a)} |\nabla^m v|^p w_{m,p}\, dx \right)$$

for $0 \leq k < m$. Furthermore,

$$\int_{B_{2R}(a)} |v|^p w\, dx \leq C \int_{B_{2R}(a)} |Eu|^p w\, dx \leq C\|u\|_{V_w^{m,p}(\Omega)}^p$$

and, by the Leibniz rule and Lemma 2.2.19 once more,

$$\int_{B_{2R}(a)} |\nabla^m v|^p w_{m,p} \, dx \le C \sum_{|\alpha|=m} \sum_{\beta \le \alpha} \frac{1}{R^{(m-|\beta|)p}} \int_{B_{2R}(a)} |D^\beta Eu|^p w_{m,p} \, dx$$

$$\le C \sum_{|\alpha|=m} \left(w(B_{2R}(a))^{-mp/N} \int_{B_{2R}(a)} |Eu|^p w \, dx \right.$$

$$\left. + \int_{B_{2R}(a)} |\nabla^m Eu|^p w_{m,p} \, dx \right)$$

$$\le C \|u\|^p_{V^{m,p}_w(\Omega)},$$

which finishes the proof. □

We finally remark that, in general, there is no relation (i.e., inclusion) between $V^{m,p}_w(\mathbf{R}^N)$ and $W^{m,p}_{w_{m,p}}(\mathbf{R}^N)$. Take for instance $N = 3$, $m = 1$, and $p = 2$. If $w(x) = |x|^2$, then $w \in A_2$, and an easy calculation shows that the function $x \mapsto (1 + |x|^2)^{-1}$ belongs to $W^{1,2}_{w_{1,2}}(\mathbf{R}^N)$ but not to $V^{1,2}_w(\mathbf{R}^N)$. Furthermore, if $w(x) = |x|^{-2}$, then $w \in A_2$ and the function $x \mapsto (1 + |x|)^{-1/2}$ belongs to $V^{1,2}_w(\mathbf{R}^N)$ but not to $W^{1,2}_{w_{1,2}}(\mathbf{R}^N)$. This also implies that there cannot be any relation between $V^{m,p}_w(\Omega)$ and $W^{m,p}_w(\Omega)$ since this, by the theorem just proved, would imply a corresponding relation between $W^{m,p}_{w_{m,p}}(\Omega)$ and $W^{m,p}_w(\Omega)$, which of course is not possible for arbitrary A_p weights w. However, if $w \in A_1$ and Ω is bounded, then $W^{m,p}_w(\Omega) \subset V^{m,p}_w(\Omega)$. Indeed, suppose that $\Omega \subset B$ for some ball B, and let $u \in W^{m,p}_w(\Omega)$. Then, by the A_1 condition,

$$\int_\Omega |\nabla^k u|^p w_{k,p} \, dx \le C \left(\frac{|B|}{w(B)} \right)^{kp/N} \int_\Omega |\nabla^k u|^p w \, dx \qquad (2.2.20)$$

for $0 \le k \le m$. We will return to some consequences of this inclusion in Section 2.4.

2.3. Hausdorff measures

In this short section, we prove some simple properties of a weighted Hausdorff measure, introduced by E. Nieminen in [91, p. 11], that will be used in the next section.

Definition 2.3.1. Let w be a given weight, and let $0 \le \alpha < N$. For a ball $B_r(a) \subset \mathbf{R}^N$, we define the *measure function* $h^{N-\alpha}_w(B_r(a))$ by

$$h^{N-\alpha}_w(B_r(a)) = \frac{w(B_r(a))}{r^\alpha}.$$

Let $0 < \rho \leq \infty$. If $E \subset \mathbf{R}^N$, we set

$$\mathcal{H}_{w,\rho}^{N-\alpha}(E) = \inf \sum_j h_w^{N-\alpha}(B_j),$$

where the infimum is taken over all countable coverings of E with open balls B_j such that each ball has radius $\leq \rho$ (when $\rho < \infty$). We then define the *weighted, $(N - \alpha)$-dimensional Hausdorff measure* of E, $\mathcal{H}_w^{N-\alpha}(E)$, as

$$\mathcal{H}_w^{N-\alpha}(E) = \lim_{\rho \to 0} \mathcal{H}_{w,\rho}^{N-\alpha}(E).$$

Remark 2.3.2. When $w = 1$, we will write $h^{N-\alpha}$ and $\mathcal{H}^{N-\alpha}$ instead of $h_1^{N-\alpha}$ and $\mathcal{H}_1^{N-\alpha}$, respectively. The measure $\mathcal{H}^{N-\alpha}$ is, up to a multiplicative constant, the usual, spherical Hausdorff measure (see Federer [38, p. 171]).

The following proposition summarizes some important properties of the measure $\mathcal{H}_w^{N-\alpha}$. Since the proofs are easy and identical to the proofs in the non-weighted case (see, e.g., the book by L. C. Evans and R. F. Gariepy [34, pp. 61–63]), we will only prove the third property as we will have occasion to refer back to this proof later.[†]

Proposition 2.3.3. *Let w be a given weight and let $0 \leq \alpha < N$.*

(a) *$\mathcal{H}_w^{N-\alpha}$ is an outer measure, i.e.,*

   (1) *$\mathcal{H}_w^{N-\alpha}(\emptyset) = 0$;*
   (2) *if $E_1 \subset E_2$, then $\mathcal{H}_w^{N-\alpha}(E_1) \leq \mathcal{H}_w^{N-\alpha}(E_2)$;*
   (3) *$\mathcal{H}_w^{N-\alpha}\left(\bigcup_{n=1}^\infty E_n\right) \leq \sum_{n=1}^\infty \mathcal{H}_w^{N-\alpha}(E_n)$.*

(b) *$\mathcal{H}_w^{N-\alpha}$ satisfies Carathéodory's criterion:[‡] if E_1 and E_2 are two subsets of $\mathbf{R}^N$ a positive distance apart, then*

$$\mathcal{H}_w^{N-\alpha}(E_1 \cup E_2) = \mathcal{H}_w^{N-\alpha}(E_1) + \mathcal{H}_w^{N-\alpha}(E_2).$$

Thus, every Borel set is $\mathcal{H}_w^{N-\alpha}$-measurable, so $\mathcal{H}_w^{N-\alpha}$ is a Borel measure.

(c) *For every subset E of $\mathbf{R}^N$, there is a Borel set B such that $B \supset E$ and $\mathcal{H}_w^{N-\alpha}(B) = \mathcal{H}_w^{N-\alpha}(E)$.*

Remark 2.3.4. For later use in Section 3.4, we remark that property (a) also holds for $\mathcal{H}_{w,\rho}^{N-\alpha}$.

[†]The fact that $\mathcal{H}_w^{N-\alpha}$ has properties (a) and (b) is also stated in Nieminen [91].

[‡]Carathéodory's criterion is the following: if μ is an outer measure on $\mathbf{R}^N$, satisfying $\mu(A \cup B) = \mu(A) + \mu(B)$ for all sets $A, B \subset \mathbf{R}^N$ with $\text{dist}(A, B) > 0$, then μ is a Borel measure; see Evans–Gariepy [34, pp. 9–11].

Proof of (c). For $k = 1, 2, \ldots$, choose coverings $\{B_j^{(k)}\}_j$ of E with balls $B_j^{(k)}$ such that each ball intersects E and has radius $\leq 1/k$, and

$$\sum_j h_w^{N-\alpha}(B_j^{(k)}) < \mathcal{H}_{w,1/k}^{N-\alpha}(E) + \frac{1}{k}.$$

Set $B = \bigcap_{k=1}^{\infty} \bigcup_j B_j^{(k)}$. Then B is Borel, in fact G_δ, and $B \supset E$. Thus, by (a), $\mathcal{H}_w^{N-\alpha}(E) \leq \mathcal{H}_w^{N-\alpha}(B)$. Note that $\mathcal{H}_{w,1/k}^{N-\alpha}(B) \leq h_w^{N-\alpha}(B)$ for every ball B. It follows that, for fixed k,

$$\mathcal{H}_{w,1/k}^{N-\alpha}(B) \leq \mathcal{H}_{w,1/k}^{N-\alpha}\left(\bigcup_j B_j^{(k)}\right) \leq \sum_j \mathcal{H}_{w,1/k}^{N-\alpha}(B_j^{(k)})$$

$$\leq \sum_j h_w^{N-\alpha}(B_j^{(k)}) < \mathcal{H}_{w,1/k}^{N-\alpha}(E) + \frac{1}{k}.$$

By letting $k \to \infty$ in the last inequality, we obtain $\mathcal{H}_w^{N-\alpha}(B) \leq \mathcal{H}_w^{N-\alpha}(E)$, whence $\mathcal{H}_w^{N-\alpha}(B) = \mathcal{H}_w^{N-\alpha}(E)$. $\square$

Remark 2.3.5. It is easy to see that the set B constructed in the proof above is included in the closure of E.

We will next show that if $w \in A_1$, then the weighted Hausdorff measure $\mathcal{H}_w^{N-\alpha}$ locally dominates the non-weighted measure $\mathcal{H}^{N-\alpha}$. This is of course natural, since an A_1 weight is locally bounded from below away from 0.

Proposition 2.3.6. *Let* $w \in A_1$, *and let* $0 \leq \alpha < N$. *Then there exists a positive constant* C, *depending only on* N *and the* A_1 *constant of* w, *such that if* E *is a subset of a ball* B, *then*

$$\mathcal{H}^{N-\alpha}(E) \leq C \frac{|B|}{w(B)} \mathcal{H}_w^{N-\alpha}(E). \tag{2.3.1}$$

Proof. Let $\varepsilon > 0$ be arbitrary, and let ρ be a positive number not greater than the radius of B. Let $\{B_j\}$ be a covering of E such that every B_j intersects E and has radius $\leq \rho$, and

$$\sum_j h_w^{N-\alpha}(B_j) < \mathcal{H}_{w,\rho}^{N-\alpha}(E) + \varepsilon. \tag{2.3.2}$$

Then $B_j \subset 3B$ for every j, so by the strong doubling property of w (Proposition 1.2.7),

$$w(B_j) \geq C \frac{|B_j|}{|B|} w(B),$$

and hence

$$h_w^{N-\alpha}(B_j) \geq C \frac{w(B)}{|B|} |B_j|^{(N-\alpha)/N}.$$

If we combine this with (2.3.2), we get

$$\mathcal{H}_{w,\rho}^{N-\alpha}(E) + \varepsilon \geq C \frac{w(B)}{|B|} \sum_j |B_j|^{(N-\alpha)/N} \geq C \frac{w(B)}{|B|} \mathcal{H}_\rho^{N-\alpha}(E),$$

and all that remains to prove (2.3.1) is to let $\rho \to 0$ and remember that ε was arbitrary. $\square$

Corollary 2.3.7. *Let $w \in A_1$ and let $0 \leq \alpha < N$. Then $\mathcal{H}^{N-\alpha}$ is absolutely continuous with respect to $\mathcal{H}_w^{N-\alpha}$.*

Proof. Suppose that $\mathcal{H}_w^{N-\alpha}(E) = 0$. Let $\mathbf{R}^N = \bigcup_j B_j$, where B_j are balls. Then, by the subadditivity of $\mathcal{H}^{N-\alpha}$ and Proposition 2.3.6,

$$\mathcal{H}^{N-\alpha}(E) = \mathcal{H}^{N-\alpha}\left(\bigcup_j B_j \cap E\right) \leq \sum_j \mathcal{H}^{N-\alpha}(B_j \cap E)$$

$$\leq C \sum_j \frac{|B_j|}{w(B_j)} \mathcal{H}_w^{N-\alpha}(B_j \cap E) = 0. \quad \square$$

2.4. Isoperimetric inequalities

In the non-weighted case, it is well-known that there is a deep connection between the isoperimetric inequality and the Sobolev type inequality for $p = 1$; in fact, the inequalities are equivalent in the sense that each inequality can be used to prove the other and the best constants are equal (see, e.g., Osserman [92, pp. 1192–1194] or Ziemer [108, pp. 81–83]). This connection, which was noticed by V. G. Maz'ya [70, pp. 884–885],[†] indicates the geometric nature of the Sobolev space theory for $p = 1$. Recently, G. David and S. Semmes showed in [30, p. 106],[‡] that if w is a continuous A_1 weight and Ω is a bounded domain in $\mathbf{R}^N$ with smooth boundary, then the following two-weighted isoperimetric inequality holds:

$$\left(\int_\Omega w\, dx\right)^{(N-1)/N} \leq C \int_{\partial\Omega} w^{(N-1)/N} \, d\mathcal{H}^{N-1}. \qquad (2.4.1)$$

Using techniques similar to those of David and Semmes, we will show in Theorem 2.4.8 that the same inequality holds for arbitrary A_1 weights, if Ω is taken

---

[†]Independently of Maz'ya, H. Federer and W. H. Fleming used similar techniques to prove the Sobolev type inequality for $p = 1$, with the best constant, starting from the classical isoperimetric inequality; see [39, p. 487].

[‡]In [30], this result is proved for smooth weights, satisfying a "strong A_∞ condition." David and Semmes also show that every A_1 weight is a strong A_∞ weight but that for every $p > 1$, there is an A_p weight, that does not satisfy the strong A_∞ condition. Conversely, the set of strong A_∞ weights is not a subset of A_p.

to be the level set $\mathcal{L}_t = \{x \in \mathbf{R}^N \,;\, |u(x)| > t\}$, $t \geq 0$, and $\partial\Omega$ is the level surface $\mathcal{E}_t = \{x \in \mathbf{R}^N \,;\, |u(x)| = t\}$, $t \geq 0$, of a smooth function u. Naturally, there will be an exceptional set of level sets and level surfaces for which (2.4.1) does not hold, stemming from the fact that the right-hand side in (2.4.1), in general, is not defined for more than almost every level surface. The inequality in Theorem 2.4.8 will then be used to prove a more general isoperimetric inequality with weighted, $(N-1)$-dimensional lower Minkowski content in the right-hand side (Theorem 2.4.12). We also show that an isoperimetric inequality holds with weighted Hausdorff measures instead of integration with respect to $w\,dx$ and $w^{(N-1)/N}\,d\mathcal{H}^{N-1}$ (Theorem 2.4.21). As corollaries to the two-weighted inequalities, we obtain corresponding single weighted isoperimetric inequalities.[†] Theorem 2.4.8 will be used in Section 2.5 to prove a Sobolev type inequality for $\overset{\circ}{V}{}_w^{m,1}(\mathbf{R}^N)$, which extends Theorem 2.2.2. This inequality in turn implies Sobolev inequalities for $\overset{\circ}{W}{}_w^{m,1}(\Omega)$, where Ω is a bounded domain in $\mathbf{R}^N$, and for $W_w^{m,1}(\Omega)$, where Ω is a bounded (ε,δ) domain in $\mathbf{R}^N$.

Another important geometric inequality with applications to analysis is W. Gustin's "boxing inequality" [49] (see also Maz'ya [76, p. 33]), which states that for every compact set $K \subset \mathbf{R}^N$, there is a covering $\{B_{r_j}(x_j)\}_{j=1}^{\infty}$ of K such that

$$\sum_{j=1}^{\infty} r_j^{N-1} \leq C\mathcal{H}^{N-1}(\partial K), \qquad (2.4.2)$$

with C only depending on N. In Theorem 2.4.23, we prove a weighted boxing inequality, that will be used in the next section.

2.4.1. Preliminary lemmas

We begin by showing a simple Poincaré type inequality, which we need for the proof of the two-weighted isoperimetric inequality.

Lemma 2.4.1. *Let $w \in A_1$ and let $B \subset \mathbf{R}^N$ be a ball. Suppose that $u \in V_w^{1,1}(B)$ and that u vanishes on a measurable subset E of B with $|E| \geq \theta|B|$, where $0 < \theta < 1$. Then*

$$\int_B |u|w\,dx \leq C\theta^{-(N-1)/N}w(B)^{1/N} \int_B |\nabla u|w^{(N-1)/N}\,dx, \qquad (2.4.3)$$

where the constant C only depends on N and the A_1 constant of w.

Remark 2.4.2. The proof of (2.4.1) by David and Semmes is based on the Wirtinger inequality

$$\int_B \left(\int_B |u(x) - u(y)|w(x)\,dx \right) w(y)\,dy$$

[†] Without giving any details, we remark that weighted isoperimetric inequalities have earlier been studied by E. Stredulinsky [100, p. 67].

$$\le Cw(B)^{(N+1)/N} \int_B |\nabla u| w^{(N-1)/N} \, dx \qquad (2.4.4)$$

for $u \in C^1(B)$, where B is a ball. The inequality (2.4.4) follows in turn from the inequality

$$|u(x) - u(y)| \le C \int_{B_{x,y}} \frac{|\nabla u(z)|}{w(B_{x,z})^{(N-1)/N}} w(z)^{(N-1)/N} \, dz$$

$$+ C \int_{B_{x,y}} \frac{|\nabla u(z)|}{w(B_{y,z})^{(N-1)/N}} w(z)^{(N-1)/N} \, dz, \qquad (2.4.5)$$

where $B_{x,y} = B_{|x-y|/2}((x+y)/2)$, and $B_{x,z}$ and $B_{y,z}$ are defined similarly. This inequality is proved in [30] through a quite complicated argument. However, it should be remarked that Wirtinger inequalities like (2.4.4) could also be proved along the same lines as in the proof below using non-weighted analogues of (2.4.5).

In the proof of Lemma 2.4.1, we use the next two lemmas. The first of these will be used frequently in later sections and can be found in Adams–Hedberg [7, pp. 55–56]). The lemma is easily proved by writing $\mu(B_r(x)) = \int_{|x-y|<r} d\mu(y)$ and changing the order of integration in the right-hand sides of the identities.

Lemma 2.4.3. *Let $\mu \in \mathcal{M}^+(\mathbf{R}^N)$, $\rho > 0$, and $0 < \alpha < N$. Then, for $x \in \mathbf{R}^N$,*

$$\int_{|x-y|<\rho} \frac{d\mu(y)}{|x-y|^{N-\alpha}} = (N-\alpha) \int_0^\rho \frac{\mu(B_r(x))}{r^{N-\alpha}} \frac{dr}{r} + \frac{\mu(B_\rho(x))}{\rho^{N-\alpha}}, \qquad (2.4.6)$$

and

$$\int_{|x-y|\ge\rho} \frac{d\mu(y)}{|x-y|^{N-\alpha}} = (N-\alpha) \int_\rho^\infty \frac{\mu(B_r(x))}{r^{N-\alpha}} \frac{dr}{r} - \frac{\mu(B_\rho(x))}{\rho^{N-\alpha}}. \qquad (2.4.7)$$

Notice that by combining (2.4.6) and (2.4.7) in the lemma, it follows that

$$\int_{\delta\le|x-y|<\rho} \frac{d\mu(y)}{|x-y|^{N-\alpha}} = (N-\alpha) \int_\delta^\rho \frac{\mu(B_r(x))}{r^{N-\alpha}} \frac{dr}{r}$$

$$+ \frac{\mu(B_\rho(x))}{\rho^{N-\alpha}} - \frac{\mu(B_\delta(x))}{\delta^{N-\alpha}}, \qquad (2.4.8)$$

whenever $0 < \delta < \rho < \infty$.

Lemma 2.4.4. *Suppose that $w \in A_1$. Then there exists a constant C, which only depends on N and the A_1 constant of w, such that if B is an open ball, then*

$$\int_B \frac{w(y)}{|x-y|^{N-1}} \, dy \le Cw(B)^{1/N} w(x)^{(N-1)/N} \qquad (2.4.9)$$

for a.e. $x \in B$.

Proof. Suppose that B has radius R. Let $x \in B$. By (2.4.6), we then have

$$\int_B \frac{w(y)}{|x-y|^{N-1}}\, dy \leq \int_{|x-y|<2R} \frac{w(y)}{|x-y|^{N-1}}\, dy$$

$$= (N-1) \int_0^{2R} \frac{w(B_t(x))}{t^{N-1}} \frac{dt}{t} + \frac{w(B_{2R}(x))}{(2R)^{N-1}}. \quad (2.4.10)$$

Using the A_1 condition (Remark 1.2.4.7), we get, for a.e. $x \in B$,

$$\int_0^{2R} \frac{w(B_t(x))}{t^{N-1}} \frac{dt}{t} = \int_0^{2R} \left(\frac{w(B_t(x))}{t^N} \right)^{(N-1)/N} w(B_t(x))^{1/N} \frac{dt}{t}$$

$$\leq C w(x)^{(N-1)/N} \int_0^{2R} w(B_t(x))^{1/N} \frac{dt}{t}. \quad (2.4.11)$$

Furthermore, then since w also belongs to A_∞ (see Proposition 1.2.9), there are positive constants C and δ such that if $0 \leq t \leq 2R$, then

$$w(B_t(x)) \leq C \left(\frac{|B_t(x)|}{|B_{2R}(x)|} \right)^\delta w(B_{2R}(x)) = C \frac{w(B_{2R}(x))}{(2R)^{\delta N}} t^{\delta N}.$$

This fact and the strong doubling property of w (Proposition 1.2.7) implies that

$$\int_0^{2R} w(B_t(x))^{1/N} \frac{dt}{t} \leq C \frac{w(B_{2R}(x))^{1/N}}{(2R)^\delta} \int_0^{2R} t^{\delta-1}\, dt \leq C w(B)^{1/N}.$$

It follows that the first term in the right-hand side of (2.4.10) is majorized by the right-hand side of (2.4.9). Finally, the last term in (2.4.10) is bounded as in (2.4.11). $\square$

Proof of Lemma 2.4.1. Since $V_w^{1,1}(B) \subset W^{1,1}(B)$, we have

$$|u(x)| \leq C\theta^{-(N-1)/N} \int_B \frac{|\nabla u(y)|}{|x-y|^{N-1}}\, dy,$$

for a.e. $x \in B$ (see Adams–Hedberg [7, p. 225]). This implies that

$$\int_B |u|w\, dx \leq C\theta^{-(N-1)/N} \int_B \left(\int_B \frac{w(x)}{|x-y|^{N-1}}\, dx \right) |\nabla u(y)|\, dy.$$

Now, according to the previous lemma,

$$\int_B \frac{w(x)}{|x-y|^{N-1}}\, dx \leq C w(B)^{1/N} w(y)^{(N-1)/N},$$

for a.e. $y \in B$, and (2.4.3) follows. $\square$

2.4.2. Extensions of some results by David and Semmes

Henceforth, we shall use the following notation: if u is a function on $\mathbf{R}^N$ and $t \geq 0$, then $\mathcal{L}_t$ and $\mathcal{E}_t$ are the level set and the level surface of $|u|$, respectively, that is

$$\mathcal{L}_t = \{x \in \mathbf{R}^N \, ; \, |u(x)| > t\}$$

and

$$\mathcal{E}_t = \{x \in \mathbf{R}^N \, ; \, |u(x)| = t\}.$$

With the aid of Lemma 2.4.1, we can now prove a local isoperimetric inequality.

Lemma 2.4.5. *Let $w \in A_1$. Then there is a positive constant C, which only depends on N and the A_1 constant of w, such that if $u \in C_0^1(\mathbf{R}^N)$, then, for every open ball B and every $t \geq 0$, satisfying $\frac{1}{4}|B| \leq |B \cap \mathcal{L}_t| \leq \frac{3}{4}|B|$,*

$$w(B \cap \mathcal{L}_t)^{(N-1)/N} \leq C \liminf_{h \to 0, \, h > 0} \frac{1}{h} \int_h^{t+h} \left(\int_{B \cap \mathcal{E}_s} w^{(N-1)/N} \, d\mathcal{H}^{N-1} \right) ds. \tag{2.4.12}$$

It follows that, for a.e. $t \geq 0$, satisfying $\frac{1}{4}|B| \leq |B \cap \mathcal{L}_t| \leq \frac{3}{4}|B|$,

$$w(B \cap \mathcal{L}_t)^{(N-1)/N} \leq C \int_{B \cap \mathcal{E}_t} w^{(N-1)/N} \, d\mathcal{H}^{N-1}. \tag{2.4.13}$$

Remark 2.4.6. The exceptional set of t, for which (2.4.13) does not hold, will, a priori, depend on the ball B.

For the proof of the lemma, we will use the following weighted co-area formula, which can be found in the book by H. Federer [38, p. 258].

Theorem 2.4.7. *Let w be a weight, and let $u \in C_0^1(\mathbf{R}^N)$. Then, if E is a measurable subset of $\mathbf{R}^N$,*

$$\int_E |\nabla u| w \, dx = \int_0^\infty \left(\int_{E \cap \mathcal{E}_t} w \, d\mathcal{H}^{N-1} \right) dt.$$

Proof of Lemma 2.4.5. Note that the inequality (2.4.13) follows from (2.4.12) using Lebesgue's differentiation theorem. Let B be a fixed open ball and let $t \geq 0$ be such that $\frac{1}{4}|B| \leq |B \cap \mathcal{L}_t| \leq \frac{3}{4}|B|$. For fixed $h > 0$, we define the function F_h by

$$F_h(x) = \frac{1}{h}(\min\{|u(x)|, t + h\} - \min\{|u(x)|, t\}),$$

for $x \in \mathbf{R}^N$, that is,

$$F_h(x) = \begin{cases} 1, & \text{if } |u(x)| \geq t + h, \\ (|u(x)| - t)/h, & \text{if } t \leq |u(x)| < t + h, \text{ and} \\ 0, & \text{if } |u(x)| < t. \end{cases}$$

Then, by the truncation theorem (Theorem 2.1.8), $F_h \in V_w^{1,1}(B)$, and

$$D_i F_h(x) = \begin{cases} D_i u(x)/h, & \text{if } t \le u(x) < t + h, \\ -D_i u(x)/h, & \text{if } -(t+h) < u(x) \le -t, \text{ and} \\ 0, & \text{otherwise} \end{cases}$$

for $i = 1, ..., N$. Let $E_h = \{x \in B \; ; \; t \le |u(x)| < t + h\}$. If we now use the Poincaré inequality in Lemma 2.4.1, we get

$$w(B \cap \mathcal{L}_{t+h}) \le \int_B |F_h| w \, dx \le C w(B)^{1/N} \frac{1}{h} \int_{E_h} |\nabla u| w^{(N-1)/N} \, dx. \quad (2.4.14)$$

An application of the co-area formula shows that

$$\int_{E_h} |\nabla u| w^{(N-1)/N} \, dx = \int_t^{t+h} \left(\int_{B \cap \mathcal{E}_s} w^{(N-1)/N} \, d\mathcal{H}^{N-1} \right) ds.$$

Hence, if we let $h \to 0$ in (2.4.14), it follows that

$$w(B \cap \mathcal{L}_t) \le C w(B)^{1/N} \liminf_{h \to 0, \, h > 0} \frac{1}{h} \int_t^{t+h} \left(\int_{B \cap \mathcal{E}_s} w^{(N-1)/N} \, d\mathcal{H}^{N-1} \right) ds.$$

Finally, by the strong doubling property of w, $w(B) \le C w(B \cap \mathcal{L}_t)$, and the lemma follows. $\square$

We can now prove the two-weighted isoperimetric inequality.

Theorem 2.4.8. *Let $w \in A_1$. Then there is a positive constant C, which only depends on N and the A_1 constant of w, such that if $u \in C_0^1(\mathbf{R}^N)$, then for a.e. $t \ge 0$,*

$$\left(\int_{\mathcal{L}_t} w \, dx \right)^{(N-1)/N} \le C \int_{\mathcal{E}_t} w^{(N-1)/N} \, d\mathcal{H}^{N-1}. \quad (2.4.15)$$

Proof. Suppose that t is such that $\mathcal{L}_t \ne \emptyset$. For every $x \in \mathcal{L}_t$, there is an open ball B_x such that $x \in B_x$ and $|B_x \cap \mathcal{L}_t| = \frac{1}{2}|B_x|$. The local isoperimetric inequality (2.4.12) implies that

$$w(B_x \cap \mathcal{L}_t)^{(N-1)/N} \le C \liminf_{h \to 0, \, h > 0} \frac{1}{h} \int_h^{t+h} \left(\int_{B_x \cap \mathcal{E}_s} w^{(N-1)/N} \, d\mathcal{H}^{N-1} \right) ds.$$

By a well-known covering lemma (see Stein [99, pp. 9–10]), we may extract a pairwise disjoint subsequence $\{B_j\}$ from the covering $\{B_x\}_{x \in \mathcal{L}_t}$ such that $\{5B_j\}$ covers $\mathcal{L}_t$. Using the elementary inequality

$$\left(\sum_j a_j \right)^s \le \sum_j a_j^s, \quad (2.4.16)$$

(which holds for $a_j \geq 0$ and $0 \leq s \leq 1$), the strong doubling property of w, Fatou's lemma, and Fubini's theorem, we then obtain

$$w(\mathcal{L}_t)^{(N-1)/N} \leq \left(\sum_j w(5B_j \cap \mathcal{L}_t) \right)^{(N-1)/N} \leq C \sum_j w(B_j \cap \mathcal{L}_t)^{(N-1)/N}$$

$$\leq C \sum_j \liminf_{h \to 0, \, h > 0} \frac{1}{h} \int_h^{t+h} \left(\int_{B_j \cap \mathcal{E}_s} w^{(N-1)/N} \, d\mathcal{H}^{N-1} \right) ds$$

$$\leq C \liminf_{h \to 0, \, h > 0} \frac{1}{h} \int_h^{t+h} \left(\sum_j \int_{B_j \cap \mathcal{E}_s} w^{(N-1)/N} \, d\mathcal{H}^{N-1} \right) ds$$

$$\leq C \liminf_{h \to 0, \, h > 0} \frac{1}{h} \int_h^{t+h} \left(\int_{\mathcal{E}_s} w^{(N-1)/N} \, d\mathcal{H}^{N-1} \right) ds.$$

The inequality (2.4.15) now follows from Lebesgue's differentiation theorem.   □

As a corollary to Theorem 2.4.8, we get the single weighted isoperimetric inequality below.

Corollary 2.4.9. *Let $w \in A_1$. Then there is a positive constant C, which only depends on N and the A_1 constant of w, such that if $u \in C_0^1(\mathbf{R}^N)$ with support in a ball B, then, for a.e. $t \geq 0$,*

$$\left(\int_{\mathcal{L}_t} w \, dx \right)^{(N-1)/N} \leq C \left(\frac{|B|}{w(B)} \right)^{1/N} \int_{\mathcal{E}_t} w \, d\mathcal{H}^{N-1}. \qquad (2.4.17)$$

In the proof of Corollary 2.4.9, we use the following well-known lemma.

Lemma 2.4.10. *Let E be a subset of $\mathbf{R}^N$ with $|E| = 0$, and let $u \in C_0^1(\mathbf{R}^N)$. Then, for a.e. $t \geq 0$, $\mathcal{H}^{N-1}(E \cap \mathcal{E}_t) = 0$.*

Proof. By the co-area formula,

$$0 = \int_E |\nabla u| \, dx = \int_0^\infty \mathcal{H}^{N-1}(E \cap \mathcal{E}_t) \, dt. \quad □$$

Proof of Corollary 2.4.9. Since $w \in A_1$,

$$w(x) \geq C \fint_B w \, dy$$

for every $x \in B$ except for x belonging to a set E of measure 0. Moreover, we know that

$$\left(\int_{\mathcal{L}_t} w \, dx \right)^{(N-1)/N} \leq C \int_{\mathcal{E}_t} w^{(N-1)/N} \, d\mathcal{H}^{N-1} \qquad (2.4.18)$$

for a.e. $t \geq 0$. Now let $t \geq 0$ be such that $\mathcal{H}^{N-1}(E \cap \mathcal{E}_t) = 0$, (2.4.18) holds, and the integral in the right-hand side of (2.4.17) is defined. Then

$$\int_{\mathcal{E}_t} w^{(N-1)/N} \, d\mathcal{H}^{N-1} \leq C \left(\frac{|B|}{w(B)} \right)^{1/N} \int_{\mathcal{E}_t} w \, d\mathcal{H}^{N-1},$$

thus proving what we wanted. □

2.4.3. Isoperimetric inequalities involving lower Minkowski content

By replacing integration with respect to $\mathcal{H}^{N-1}$ in Theorem 2.4.8 with lower, $(N-1)$-dimensional Minkowski content, it is possible to obtain isoperimetric inequalities that hold for arbitrary compact sets. For the definition of Minkowski content in the non-weighted case, see Federer [38, p. 273]. Notice in passing that the right-hand side in the local isoperimetric inequality (2.4.12) is a quantity similar to the Minkowski content.

Definition 2.4.11. Let w be a weight. We then define the *weighted, $(N-1)$-dimensional lower Minkowski content* $\mathcal{M}^{N-1}_{w,*}(E)$ of a measurable set $E \subset \mathbf{R}^N$ by

$$\mathcal{M}^{N-1}_{w,*}(E) = \liminf_{h \to 0, \, h > 0} \frac{w(E_h)}{h},$$

where E_h is the set $\{x \in \mathbf{R}^N \, ; \, \mathrm{dist}(x, E) < h\}$.

The proof of the following theorem is adapted from Federer [38, p. 504].

Theorem 2.4.12. *Let $w \in A_1$. Then there is a positive constant C, which only depends on N and the A_1 constant of w, such that, for every compact subset K of $\mathbf{R}^N$,*

$$w(K)^{(N-1)/N} \leq C \mathcal{M}^{N-1}_{w_{1,1},*}(\partial K), \tag{2.4.19}$$

where $w_{1,1} = w^{(N-1)/N}$.

Proof. Let $\varphi \in C_0^\infty(\mathbf{R}^N)$ be such that $\mathrm{supp}\, \varphi \subset B_{1/2}(0)$ and $\int_{\mathbf{R}^N} \varphi \, dx = 1$. For $h > 0$, set $\varphi_h(x) = h^{-N} \varphi(x/h)$, $x \in \mathbf{R}^N$, and $\chi_h = \varphi_h * \chi_K$. Let $\mathcal{L}_t^h$ and $\mathcal{E}_t^h$ be the level set and the level surface of χ_h, respectively. Note that if $\mathrm{dist}(x, \partial K) \geq h$, then

$$\chi_h(x) = \begin{cases} 1, & \text{if } x \in K, \\ 0, & \text{if } x \notin K, \end{cases}$$

so $D_i \chi_h(x) = 0$, $i = 1, \dots, N$. Also, for arbitrary x,

$$|D_i \varphi_h| * \chi_K(x) \leq \frac{1}{h} \int_{\mathbf{R}^N} |D_i \varphi| \, dx.$$

Using this, we find

$$\int_0^\infty \left(\int_{\mathcal{E}_s^h} w^{(N-1)/N} \, d\mathcal{H}^{N-1} \right) ds = \int_{(\partial K)_h} |\nabla \chi_h| w^{(N-1)/N} \, dx$$

$$\leq \sqrt{N} \frac{w_{1,1}((\partial K)_h)}{h} \sum_{i=1}^N \int_{\mathbf{R}^N} |D_i \varphi| \, dx$$

$$= C \frac{w_{1,1}((\partial K)_h)}{h}.$$

It follows that the set of t, $0 \leq t \leq \frac{1}{2}$, such that

$$\int_{\mathcal{E}_t^h} w^{(N-1)/N} \, d\mathcal{H}^{N-1} \leq 2C \frac{w_{1,1}((\partial K)_h)}{h} \tag{2.4.20}$$

has positive measure. Let t, $0 \leq t \leq \frac{1}{2}$, be such that (2.4.20) and the isoperimetric inequality (2.4.15) hold. Then

$$w(\mathcal{L}_{1/2}^h)^{(N-1)/N} \leq w(\mathcal{L}_t^h)^{(N-1)/N} \leq C \int_{\mathcal{E}_t^h} w^{(N-1)/N} \, d\mathcal{H}^{N-1}$$

$$\leq C \frac{w_{1,1}((\partial K)_h)}{h}. \tag{2.4.21}$$

We know that $\varphi_h * \chi_K \to \chi_K$ a.e., as $h \to 0$, and it is easy to see that this implies that $w(\mathcal{L}_{1/2}^h) \to w(K)$. Letting $h \to 0$ in (2.4.21), we thus obtain (2.4.19). $\square$

Remark 2.4.13. It is not so difficult to see that if w is sufficiently regular, say continuous, and ∂K is smooth, then

$$\mathcal{M}_{w_{1,1},*}^{N-1}(\partial K) = \int_{\partial K} w^{(N-1)/N} \, d\mathcal{H}^{N-1};$$

cf. Ziemer [108, pp. 82–83].

The corollary below to Theorem 2.4.12 is proved in a similar manner as Corollary 2.4.9.

Corollary 2.4.14. Let $w \in A_1$. Then there is a positive constant C, which only depends on N and the A_1 constant of w, such that if K is a compact subset of a ball B, then

$$w(K)^{(N-1)/N} \leq C \left(\frac{|B|}{w(B)} \right)^{1/N} \mathcal{M}_{w,*}^{N-1}(\partial K).$$

2.4.4. Isoperimetric inequalities with Hausdorff measures

Our next aim is to show how one can replace the left- and right-hand sides in the isoperimetric inequality (2.4.15) with weighted Hausdorff measures. The idea is simple: we prove that $\mathcal{H}^{N-1}_{w_{1,1}}|_{\mathcal{E}_t}$, the restriction of $\mathcal{H}^{N-1}_{w_{1,1}}$ to a level surface $\mathcal{E}_t$, is absolutely continuous with respect to $\mathcal{H}^{N-1}|_{\mathcal{E}_t}$ with Radon–Nikodym derivative comparable to $w^{(N-1)/N}$ for a.e. $t \geq 0$. Similarly, the Radon–Nikodym derivative of $\mathcal{H}^N_w$ with respect to $\mathcal{H}^N$ will be shown to be comparable to w.

Lemma 2.4.15. *Let $w \in A_1$. Then there is a positive constant C, which only depends on N and the A_1 constant of w, such that if $u \in C^N_0(\mathbf{R}^N)$, then for a.e. $t \geq 0$ and for every compact set $K \subset \mathbf{R}^N$,*

$$\mathcal{H}^{N-1}_{w_{1,1}}(K \cap \mathcal{E}_t) \leq C \int_{K \cap \mathcal{E}_t} w^{(N-1)/N} \, d\mathcal{H}^{N-1}. \qquad (2.4.22)$$

Proof. We denote by $\mathcal{B}$ the family of balls in $\mathbf{R}^N$ with rational centers and rational radii. Let E be the set of nonnegative t such that, for every $B \in \mathcal{B}$, $|B \cap \mathcal{L}_t| < \frac{1}{4}|B|$, or $|B \cap \mathcal{L}_t| > \frac{3}{4}|B|$, or $\frac{1}{4}|B| \leq |B \cap \mathcal{L}_t| \leq \frac{3}{4}|B|$ and the second inequality in Lemma 2.4.5 holds. Then the complement of E with respect to $\mathbf{R}_+$ has measure 0. Let $t \in E$ such that $\mathcal{E}_t$ equals $\partial \mathcal{L}_t$ and is a smooth, $(N-1)$-dimensional manifold. By a theorem of A. P. Morse [84] (see also Maz'ya [76, p. 35]), this holds for a.e. $t \geq 0$. Furthermore, suppose that the integral $\int_{\mathcal{E}_t} w^{(N-1)/N} \, d\mathcal{H}^{N-1}$ is finite. Define $G_\varepsilon = \{x \in \mathbf{R}^N \; ; \, \mathrm{dist}(x, K) < \varepsilon\}$, $\varepsilon > 0$. Then, by monotone convergence,

$$\int_{G_\varepsilon \cap \mathcal{E}_t} w^{(N-1)/N} \, d\mathcal{H}^{N-1} \to \int_{K \cap \mathcal{E}_t} w^{(N-1)/N} \, d\mathcal{H}^{N-1}, \quad \text{as } \varepsilon \to 0.$$

Let $\varepsilon > 0$ be fixed, and let $0 < \rho < \frac{1}{2}\varepsilon$. If $x \in K \cap \mathcal{E}_t$, then since

$$\lim_{r \to 0} \frac{|B_r(x) \cap \mathcal{L}_t|}{|B_r(x)|} = \frac{1}{2},$$

there is a rational ball B_x, containing x, such that the radius of B_x is $\leq \rho$ and $\frac{1}{4}|B_x| \leq |B_x \cap \mathcal{L}_t| \leq \frac{3}{4}|B_x|$. We then extract a pairwise disjoint subsequence $\{B_j\}$ from the covering $\{B_x\}_{x \in \mathcal{E}_t}$ such that $\{5B_j\}$ covers $K \cap \mathcal{E}_t$. Let r_j be the radius of B_j. Using Hölder's inequality, the strong doubling property of w, and Lemma 2.4.5, we obtain

$$\mathcal{H}^{N-1}_{w_{1,1},5\rho}(K \cap \mathcal{E}_t) \leq \sum_j \frac{w_{1,1}(5B_j)}{5r_j}$$

$$\leq C \sum_j w(5B_j)^{(N-1)/N}$$

$$\leq C \sum_j w(B_j \cap \mathcal{L}_t)^{(N-1)/N}$$

$$\leq C \sum_j \int_{B_j \cap \mathcal{E}_t} w^{(N-1)/N} \, d\mathcal{H}^{N-1}$$

$$\leq C \int_{G_\epsilon \cap \mathcal{E}_t} w^{(N-1)/N} \, d\mathcal{H}^{N-1}.$$

Hence

$$\mathcal{H}^{N-1}_{w_{1,1}, 5\rho}(K \cap \mathcal{E}_t) \leq C \int_{G_\epsilon \cap \mathcal{E}_t} w^{(N-1)/N} \, d\mathcal{H}^{N-1}$$

for $0 < \rho < \frac{1}{2}\epsilon$. If we finally let $\rho \to 0$ and then $\epsilon \to 0$ in the last inequality, we arrive at (2.4.22). $\square$

Given $u \in C_0^N(\mathbf{R}^N)$, consider the measure $\mathcal{H}^{N-1}_{w_{1,1}}|_{\mathcal{E}_t}$. It is an outer Borel measure because the same holds for $\mathcal{H}^{N-1}_{w_{1,1}}$ and $\mathcal{E}_t$ is a Borel set (in fact, closed). Moreover, by Theorem 2.4.7 and Lemma 2.4.15, $\mathcal{H}^{N-1}_{w_{1,1}}|_{\mathcal{E}_t}$ is finite for a.e. $t \geq 0$ and thus regular.

Corollary 2.4.16. *Let $w \in A_1$. Then there is a positive constant C, which only depends on N and the A_1 constant of w, such that if $u \in C_0^N(\mathbf{R}^N)$, then for a.e. $t \geq 0$ and for every Borel set $E \subset \mathbf{R}^N$,*

$$\mathcal{H}^{N-1}_{w_{1,1}}(E \cap \mathcal{E}_t) \leq C \int_{E \cap \mathcal{E}_t} w^{(N-1)/N} \, d\mathcal{H}^{N-1}. \tag{2.4.23}$$

Proof. By the inner regularity of $\mathcal{H}^{N-1}_{w_{1,1}}|_{\mathcal{E}_t}$ and Lemma 2.4.15, we have

$$\mathcal{H}^{N-1}_{w_{1,1}}(E \cap \mathcal{E}_t) = \sup \mathcal{H}^{N-1}_{w_{1,1}}(K \cap \mathcal{E}_t) \leq \sup \int_{K \cap \mathcal{E}_t} w^{(N-1)/N} \, d\mathcal{H}^{N-1}$$

$$\leq C \int_{E \cap \mathcal{E}_t} w^{(N-1)/N} \, d\mathcal{H}^{N-1},$$

where the supremum is over all compact subsets K of E. $\square$

Corollary 2.4.17. *Let $w \in A_1$, and let $u \in C_0^N(\mathbf{R}^N)$. Then, for a.e. $t \geq 0$, $\mathcal{H}^{N-1}_{w_{1,1}}|_{\mathcal{E}_t}$ is absolutely continuous with respect to $\mathcal{H}^{N-1}|_{\mathcal{E}_t}$.*

We now claim that $\mathcal{H}^{N-1}_{w_{1,1}}|_{\mathcal{E}_t}$ is a Radon measure for a.e. $t \geq 0$. To prove this, we only have to show that, for every $E \subset \mathbf{R}^N$ there is a Borel set $B \supset E$ such that $\mathcal{H}^{N-1}_{w_{1,1}}|_{\mathcal{E}_t}(B) = \mathcal{H}^{N-1}_{w_{1,1}}|_{\mathcal{E}_t}(E)$. But, by property (c) in Lemma 2.3.3, there exists a Borel set $B \supset E \cap \mathcal{E}_t$ such that $\mathcal{H}^{N-1}_{w_{1,1}}(B) = \mathcal{H}^{N-1}_{w_{1,1}}(E \cap \mathcal{E}_t) = \mathcal{H}^{N-1}_{w_{1,1}}|_{\mathcal{E}_t}(E)$, and, by the remark immediately after Lemma 2.3.3, B is included in the closure of $E \cap \mathcal{E}_t$. It follows that $B \subset \mathcal{E}_t$ (since $\overline{E \cap \mathcal{E}_t} \subset \overline{\mathcal{E}_t} = \mathcal{E}_t$), and thus, $\mathcal{H}^{N-1}_{w_{1,1}}(B) = \mathcal{H}^{N-1}_{w_{1,1}}(B \cap \mathcal{E}_t) = \mathcal{H}^{N-1}_{w_{1,1}}|_{\mathcal{E}_t}(B)$, so $\mathcal{H}^{N-1}_{w_{1,1}}|_{\mathcal{E}_t}(B) = \mathcal{H}^{N-1}_{w_{1,1}}|_{\mathcal{E}_t}(E)$.

Proposition 2.4.18. *Let $w \in A_1$. Then there are positive constants C_1 and C_2, which only depend on N and the A_1 constant of w, such that if $u \in C_0^N(\mathbf{R}^N)$, then, for a.e. $t \geq 0$,*

$$C_1 w(x)^{(N-1)/N} \leq \frac{d\mathcal{H}^{N-1}_{w_{1,1}}|_{\mathcal{E}_t}}{d\mathcal{H}^{N-1}|_{\mathcal{E}_t}}(x) \leq C_2 w(x)^{(N-1)/N} \tag{2.4.24}$$

$\mathcal{H}^{N-1}$-a.e. on $\mathcal{E}_t$. We here use the notation

$$\frac{d\mathcal{H}^{N-1}_{w_{1,1}}|_{\mathcal{E}_t}}{d\mathcal{H}^{N-1}|_{\mathcal{E}_t}}$$

for the Radon–Nikodym derivative of $\mathcal{H}^{N-1}_{w_{1,1}}|_{\mathcal{E}_t}$ with respect to $\mathcal{H}^{N-1}|_{\mathcal{E}_t}$.

Proof. We know that, for a.e. $t \geq 0$,

$$\frac{d\mathcal{H}^{N-1}_{w_{1,1}}|_{\mathcal{E}_t}}{d\mathcal{H}^{N-1}|_{\mathcal{E}_t}}(x) = \lim_{r \to 0} \frac{\mathcal{H}^{N-1}_{w_{1,1}}|_{\mathcal{E}_t}(B_r(x))}{\mathcal{H}^{N-1}|_{\mathcal{E}_t}(B_r(x))}$$

$\mathcal{H}^{N-1}$-a.e. on $\mathcal{E}_t$; this follows from the Radon–Nikodym theorem and the Lebesgue–Besicovitch differentiation theorem (see Ziemer [108, p. 40]). Let $x \in \mathcal{E}_t$. Then, by Lemma 2.2.17 and Lemma 2.3.6,

$$\frac{\mathcal{H}^{N-1}_{w_{1,1}}|_{\mathcal{E}_t}(B_r(x))}{\mathcal{H}^{N-1}|_{\mathcal{E}_t}(B_r(x))} \geq C \frac{w_{1,1}(B_r(x))}{|B_r(x)|} \geq C \left(\frac{w(B_r(x))}{|B_r(x)|}\right)^{(N-1)/N}.$$

As $r \to 0$, the right-hand side of this inequality tends to $w(x)^{(N-1)/N}$ a.e. and thus, for a.e. $t \geq 0$, $\mathcal{H}^{N-1}$-a.e. on $\mathcal{E}_t$ (Lemma 2.4.10). This proves the first inequality in (2.4.24). The second inequality follows from Corollary 2.4.16. We have

$$\frac{\mathcal{H}^{N-1}_{w_{1,1}}|_{\mathcal{E}_t}(B_r(x))}{\mathcal{H}^{N-1}|_{\mathcal{E}_t}(B_r(x))} = \frac{\mathcal{H}^{N-1}_{w_{1,1}}|_{\mathcal{E}_t}(B_r(x))}{\int_{B_r(x) \cap \mathcal{E}_t} w^{(N-1)/N} \, d\mathcal{H}^{N-1}} \fint_{B_r(x) \cap \mathcal{E}_t} w^{(N-1)/N} \, d\mathcal{H}^{N-1}$$

$$\leq C \fint_{B_r(x) \cap \mathcal{E}_t} w^{(N-1)/N} \, d\mathcal{H}^{N-1},$$

where we temporarily use the notation

$$\fint_{B_r(x) \cap \mathcal{E}_t} w^{(N-1)/N} \, d\mathcal{H}^{N-1} = \mathcal{H}^{N-1}(\mathcal{E}_t \cap B_r(x))^{-1} \int_{B_r(x) \cap \mathcal{E}_t} w^{(N-1)/N} \, d\mathcal{H}^{N-1}.$$

By Lebesgue's differentiation theorem, the last integral tends to $w(x)^{(N-1)/N}$ $\mathcal{H}^{N-1}$-a.e. on $\mathcal{E}_t$, as $r \to 0$ (assuming that $\mathcal{E}_t$ is smooth). □

We now prove that the Radon–Nikodym derivative of $\mathcal{H}^N_w$ with respect to $\mathcal{H}^N$ is comparable to w.

Lemma 2.4.19. *Let $w \in A_1$. Then there is a positive constant C, which only depends on N and the A_1 constant of w, such that, for a.e. $t \geq 0$ and for every Borel set $E \subset \mathbf{R}^N$,*

$$\mathcal{H}_w^N(E) \leq C \int_E w \, dx.$$

Proof. The proof is very similar to that of Lemma 2.4.15 but somewhat simpler. We may assume that E is compact. As before, we let G_ε be an ε-neighbourhood of E, and let $0 < \rho < \frac{1}{2}\varepsilon$. From the covering $\{B_\rho(x)\}_{x \in E}$ we extract a pairwise disjoint subsequence $\{B_j\}$ such that $\{5B_j\}$ covers E. We get

$$\mathcal{H}_{w,5\rho}^N(\mathcal{E}_t) \leq \sum_j w(5B_j) \leq C \sum_j w(B_j) \leq C \int_{G_\varepsilon} w \, dx,$$

and then finish by first letting $\rho \to 0$ and then $\varepsilon \to 0$. $\quad\square$

Proposition 2.4.20. *Let $w \in A_1$. Then there are positive constants C_1 and C_2, which only depend on N and the A_1 constant of w, such that if $u \in C_0^N(\mathbf{R}^N)$, then for a.e. $x \in \mathbf{R}^N$,*

$$C_1 w(x) \leq \frac{d\mathcal{H}_w^N}{d\mathcal{H}^N}(x) \leq C_2 w(x).$$

Proof. By Lemma 2.3.6 and Lemma 2.4.19, we have

$$C_1 \frac{w(B_r(x))}{|B_r(x)|} \lesssim \frac{\mathcal{H}_w^N(B_r(x))}{\mathcal{H}^N(B_r(x))} = \frac{\mathcal{H}_w^N(B_r(x))}{w(B_r(x))} \fint_{B_r(x)} w \, dy \leq C_2 \frac{w(B_r(x))}{|B_r(x)|},$$

and the theorem follows as before. $\quad\square$

By combining Theorem 2.4.8 with Propositions 2.4.18 and 2.4.20, we get a two-weighted isoperimetric inequality for Hausdorff measures. This inequality also implies a corresponding single weighted inequality.

Theorem 2.4.21. *Let $w \in A_1$. Then there is a positive constant C, which only depends on N and the A_1 constant of w, such that if $u \in C_0^N(\mathbf{R}^N)$, then for a.e. $t \geq 0$,*

$$\mathcal{H}_w^N(\mathcal{L}_t)^{(N-1)/N} \leq C\mathcal{H}_{w_{1,1}}^{N-1}(\partial\mathcal{L}_t).$$

Corollary 2.4.22. *Let $w \in A_1$. Then there is a positive constant C, which only depends on N and the A_1 constant of w, such that if $u \in C_0^N(\mathbf{R}^N)$ with support in a ball B, then for a.e. $t \geq 0$,*

$$\mathcal{H}_w^N(\mathcal{L}_t)^{(N-1)/N} \leq C \left(\frac{|B|}{w(B)}\right)^{1/N} \mathcal{H}_w^{N-1}(\partial\mathcal{L}_t). \tag{2.4.25}$$

Proof. Suppose that the radius of B is R, and let $\{B_j\}$ be a covering of $\partial\mathcal{L}_t$ such that each ball B_j has radius $r_j \leq R$. Then $B_j \subset 3B$ for every j. It now follows from Hölder's inequality and the strong doubling property of w that

$$\frac{w_{1,1}(B_j)}{r_j} \leq Cw(B_j)^{(N-1)/N} \leq C\left(\frac{|B|}{w(B)}\right)^{1/N} \frac{w(B_j)}{r_j},$$

from which we easily get (2.4.25). □

2.4.5. A boxing inequality

We will finally prove a weighted boxing inequality.

Theorem 2.4.23. *Let $w \in A_1$, and let $u \in C_0^N(\mathbf{R}^N)$. Then, for a.e. $t \geq 0$, there exists a countable covering $\{B_j\}$ of $\mathcal{L}_t$, where each B_j is an open ball with radius r_j, such that*

$$\sum_j \frac{w(B_j)}{r_j} \leq C \int_{\mathcal{E}_t} w \, d\mathcal{H}^{N-1}, \tag{2.4.26}$$

and the constant C only depends on N and the A_1 constant of w. Moreover, if the support of u is contained in some ball B, then the balls B_j may be chosen to have their centers in B and their radii not greater than the radius of B.

Lemma 2.4.24. *Let $w \in A_1$. Then there is a constant $C > 0$, which only depends on N and the A_1 constant of w, such that if $u \in C_0^N(\mathbf{R}^N)$, then for every ball B with radius r and for a.e. $t \geq 0$ such that $\frac{1}{4}|B| \leq |B \cap \mathcal{L}_t| \leq \frac{3}{4}|B|$, we have*

$$\frac{w(B)}{r} \leq C \int_{B \cap \mathcal{E}_t} w \, d\mathcal{H}^{N-1}. \tag{2.4.27}$$

Proof. This follows from Lemma 2.4.5 (the inequality (2.4.13)) using the same type of argument as in the proof of Corollary 2.4.9. □

Proof of Theorem 2.4.23. The proof is essentially the same as that of Theorem 2.4.8. One first shows that, for a.e. $t \geq 0$ such that $\mathcal{L}_t \neq \emptyset$, there is a family $\{B_j\}$ of pairwise disjoint, rational balls B_j, with centers in $\mathcal{L}_t$, such that $\{5B_j\}$ covers $\mathcal{L}_t$, $\frac{1}{4}|B_j| \leq |B_j \cap \mathcal{L}_t| \leq \frac{3}{4}|B_j|$, and (2.4.27) holds for every B_j. The inequality (2.4.26) then follows in a now familiar manner. Finally, if r_j is the radius of B_j, then, by the doubling property of w,

$$\frac{w(5B_j)}{5r_j} \leq C \frac{w(B_j)}{r_j},$$

for every j, so the balls $\{5B_j\}$ have the required properties.

Now suppose $\operatorname{supp} u \subset B$. We will show that the covering $\{5B_j\}$, if necessary, can be modified so that every ball has radius $\leq$ the radius of B. It is easy to see that every B_j has radius $\leq$ twice the radius of B. Indeed, if this was not the case, then we would have $\mathcal{L}_t \subset B_j$ for some j, and hence

$$\tfrac{1}{4}|B_j| \leq |B_j \cap \mathcal{L}_t| = |\mathcal{L}_t| \leq |B| < 2^{-N}|B_j| \leq \tfrac{1}{4}|B_j|,$$

which is a contradiction. But if the radius of some B_j is greater than the radius of B, then we may replace the covering $\{5B_j\}$ with the single ball B. In fact, $B \supset \mathcal{L}_t$, and if R is the radius of B, then

$$\frac{w(B)}{R} \leq \frac{w(3B)}{R} \leq CR^{-1}\frac{|3B|}{|B_j|}w(B_j) \leq C\frac{w(B_j)}{r_j} \leq C\int_{\mathcal{E}_t} w \, d\mathcal{H}^{N-1}.$$

This completes the proof. $\square$

2.5. Some Sobolev type inequalities

The isoperimetric inequality (2.4.15) in the previous section is used below to establish a Sobolev type inequality for the space $\overset{\circ}{V}{}_w^{m,1}(\mathbf{R}^N)$. Consequences of this result are corresponding inequalities for $\overset{\circ}{W}{}_w^{m,1}(\Omega)$, where Ω is bounded, and $W_w^{m,1}(\Omega)$, where Ω is a bounded (ε, δ) domain.

Theorem 2.5.1. *Suppose that $w \in A_1$. Let k and m be integers such that $0 \leq k < m < N$, and set $1/1^* = 1 - (m-k)/N$. Then there exists a positive constant C, which only depends on k, m, N, and the A_1 constant of w, such that if $u \in \overset{\circ}{V}{}_w^{m,1}(\mathbf{R}^N)$, then*

$$\left(\int_{\mathbf{R}^N} |\nabla^k u|^{1^*} w_{k,1^*} \, dx\right)^{1/1^*} \leq C \int_{\mathbf{R}^N} |\nabla^m u| w_{m,1} \, dx, \tag{2.5.1}$$

where $w_{k,1^*} = w^{(N-k1^*)/N}$ and $w_{m,1} = w^{(N-m)/N}$ as before.

Proof. It suffices to prove the inequality for functions $u \in C_0^\infty(\mathbf{R}^N)$ and we may assume that $k = 0$. Let us first consider the case $m = 1$. For $t \geq 0$, set

$$u_t(x) = \begin{cases} u(x), & \text{if } |u(x)| \leq t, \text{ and} \\ t, & \text{if } |u(x)| > t \end{cases}$$

and

$$f_t(x) = \left(\int_{\mathbf{R}^N} |u_t|^{\frac{N}{N-1}} w \, dx\right)^{(N-1)/N}.$$

Then $|u_{t+h}(x)| \leq |u_t(x)| + h\chi_{\mathcal{L}_t}(x)$, which implies that

$$f(t+h) \leq f(t) + hw(\mathcal{L}_t)^{(N-1)/N}.$$

Thus, $f'(t) \leq w(\mathcal{L}_t)^{(N-1)/N}$ a.e. Theorem 2.4.9 and the co-area formula then shows that

$$\left(\int_\Omega |u|^{\frac{N}{N-1}} w \, dx \right)^{(N-1)/N} = \int_0^\infty f'(t) \, dt \leq \int_0^\infty w(\mathcal{L}_t)^{(N-1)/N} \, dt$$

$$\leq C \left(\frac{|B|}{w(B)} \right)^{1/N} \int_0^\infty \left(\int_{\mathcal{E}_t} w \, d\mathcal{H}^{N-1} \right) dt$$

$$= C \left(\frac{|B|}{w(B)} \right)^{1/N} \int_\Omega |\nabla u| w \, dx.$$

The general case follows by induction. Assume that (2.5.1) holds for $1 \leq m < N$, and let $m + 1 < N$. If we write $N/(N - (m+1)) = qN/(N-1)$, where $q = (N-1)/(N-(m+1))$, and apply (2.5.1) for one derivative, we find

$$\left(\int_{\mathbf{R}^N} |u|^{N/(N-(m+1))} w \, dx \right)^{(N-(m+1))/N} \leq C \int_{\mathbf{R}^N} |u|^{q-1} |\nabla u| w^{(N-1)/N} \, dx. \tag{2.5.2}$$

We now split $w^{(N-1)/N}$ into $w^{m/N} w^{(N-(m+1))/N}$, and then use Hölder's inequality with exponent $N/(N-m)$ and (2.5.1) for m derivatives to obtain

$$\int_{\mathbf{R}^N} |u|^{q-1} |\nabla u| w^{(N-1)/N} \, dx$$

$$\leq \left(\int_{\mathbf{R}^N} |u|^{N/(N-(m+1))} w \, dx \right)^{m/N}$$

$$\times \left(\int_{\mathbf{R}^N} |\nabla u|^{N/(N-m)} w^{(N-(m+1))/(N-m)} \, dx \right)^{(N-m)/N}$$

$$\leq C \left(\int_{\mathbf{R}^N} |u|^{N/(N-(m+1))} w \, dx \right)^{m/N} \int_{\mathbf{R}^N} |\nabla^{m+1} u| w^{(N-(m+1))/N} \, dx.$$

By inserting this into (2.5.2), we get (2.5.1) for $m + 1$. $\square$

Theorem 2.5.2. *Let Ω be a bounded domain in $\mathbf{R}^N$, and let B be a ball containing Ω. Let $w \in A_1$, and suppose that $u \in \overset{\circ}{W}^{m,1}_w(\Omega)$.*

(a) *If $1 \leq m < N$, then*

$$\left(\int_\Omega |u|^{N/(N-m)} w \, dx \right)^{(N-m)/N} \leq C \left(\frac{|B|}{w(B)} \right)^{m/N} \int_\Omega |\nabla^m u| w \, dx. \tag{2.5.3}$$

(b) *If $m \geq N$, then $u \in C(\Omega)$ and*

$$\sup_{x \in \Omega} |u(x)| \leq C \frac{|B|^{m/N}}{w(B)} \int_\Omega |\nabla^m u| w \, dx. \tag{2.5.4}$$

The constants appearing in (2.5.3) and (2.5.4) depend only on m, N, and the A_1 constant of w.[†]

Proof. (a) This follows from Theorem 2.5.1 as in the inequality (2.2.20).

(b) It is enough to prove (2.5.4) for functions $u \in C_0^\infty(\Omega)$. First, suppose that $m = N$. For $x = (x_1, \ldots, x_N) \in \mathbf{R}^N$, we have

$$u(x) = \int_{-\infty}^{x_1} \cdots \int_{-\infty}^{x_N} \frac{\partial^N u(y)}{\partial y_1 \ldots \partial y_N} \, dy_N \ldots dy_1.$$

Remark 1.2.4.1 then implies that

$$|u(x)| \leq \int_\Omega |\nabla^N u| \, dy \leq C \frac{|B|}{w(B)} \int_\Omega |\nabla^N u| w \, dy,$$

which is (2.5.4) for $m = N$. When $m > N$, we use the fact that

$$|\nabla^N u(y)| \leq C \int_\Omega \frac{|\nabla^m u(z)|}{|y - z|^{N-(m-N)}} \, dz$$

for $y \in B$; see (2.2.8). Hence

$$\int_B |\nabla^N u| w \, dy \leq C \int_\Omega \left(\int_B \frac{w(y)}{|y - z|^{N-(m-N)}} \, dy \right) |\nabla^m u(z)| \, dz.$$

But if R is the radius of B and $z \in B$, then

$$\int_B \frac{w(y)}{|y - z|^{N-(m-N)}} \, dy \leq \int_{|y-z|<2R} \frac{w(y)}{|y - z|^{N-(m-N)}} \, dy \leq C R^{m-N} M w(z)$$

(here we use (2.1.10)), and since $w \in A_1$, $Mw(z) \leq Cw(z)$ a.e. in B. We have thus proved that

$$|u(x)| \leq C \frac{|B|}{w(B)} R^{m-N} \int_\Omega |\nabla^m u| w \, dz = C \frac{|B|^{m/N}}{w(B)} \int_\Omega |\nabla^m u| w \, dz. \quad \square$$

Techniques similar to those in Theorem 2.2.20 gives the following corollary.

Corollary 2.5.3. *Suppose that $w \in A_1$, and let $m \geq 1$ be an integer. Let Ω be a bounded (ε, δ) domain in $\mathbf{R}^N$, and suppose that $u \in W_w^{m,1}(\Omega)$.*

(a) If $1 \leq m < N$, then

$$\left(\int_\Omega |u|^{N/(N-m)} w \, dx \right)^{(N-m)/N} \leq C \|u\|_{W_w^{m,1}(\Omega)}. \tag{2.5.5}$$

[†]Corresponding results for the case $1 < p < \infty$ can be found in Fabes–Kenig–Serapioni [36] and Miller [82].

(b) *If $m \geq N$, then $u \in C(\Omega)$ and*

$$\sup_{x \in \Omega} |u(x)| \leq C\|u\|_{W_w^{m,1}(\Omega)}. \tag{2.5.6}$$

The constants appearing in (2.5.5) and (2.5.6) depend only on Ω, m, N, and the A_1 constant of w.

Theorem 2.5.2 has the following corollary, which gives a simple characterization of those A_1 weights w for which there is a Sobolev type inequality for $W_w^{m,1}(\mathbf{R}^N)$.

Corollary 2.5.4. *Let $w \in A_1$, and let m be an integer, $1 \leq m < N$. If there is a constant $a > 0$ such that $w(x) \geq a$ for a.e. $x \in \mathbf{R}^N$, then*

$$\left(\int_{\mathbf{R}^N} |u|^{N/(N-m)} w \, dx\right)^{(N-m)/N} \leq Ca^{-m/N} \int_{\mathbf{R}^N} |\nabla^m u| w \, dx \tag{2.5.7}$$

for every $u \in W_w^{m,1}(\mathbf{R}^N)$, where C only depends on m, N and the A_1 constant of w. Conversely, if (2.5.7) holds for every $u \in W_w^{m,1}(\mathbf{R}^N)$, then $w(x) \geq Ca$ for a.e. $x \in \mathbf{R}^N$, with C as before.

Proof. First, suppose that $w(x) \geq a > 0$ a.e. It suffices to prove that (2.5.7) holds for every $u \in C_0^\infty(\mathbf{R}^N)$. Let $u \in C_0^\infty(\mathbf{R}^N)$, with support in a ball B. Then, by Theorem 2.5.2,

$$\left(\int_{\mathbf{R}^N} |u|^{N/(N-m)} w \, dx\right)^{(N-m)/N} \leq C\left(\frac{|B|}{w(B)}\right)^{m/N} \int_{\mathbf{R}^N} |\nabla^m u| w \, dx. \tag{2.5.8}$$

But $w(B) \geq a|B|$, and this together with (2.5.8) gives (2.5.7).

For the converse, let $\varphi \in C_0^\infty(\mathbf{R}^N)$ with support in $B_2(0)$ be such that $\varphi = 1$ on $\overline{B_1(0)}$. If we then apply (2.5.7) to the function $\psi(x) = \varphi((x-x_0)/r)$, $x \in \mathbf{R}^N$, where $x_0 \in \mathbf{R}^N$ and $r > 0$ are arbitrary, we get

$$\begin{aligned}
w(B_r(x_0))^{(N-m)/N} &\leq \left(\int_{\mathbf{R}^N} |\psi|^{N/(N-m)} w \, dx\right)^{(N-m)/N} \\
&\leq Ca^{-m/N} \int_{\mathbf{R}^N} |\nabla^m \psi| w \, dx \\
&\leq Ca^{-m/N} r^{-m} w(B_{2r}(x_0)) \\
&\leq Ca^{-m/N} r^{-m} w(B_r(x_0)),
\end{aligned}$$

or, equivalently,

$$\fint_{B_r(x_0)} w \, dy \geq Ca.$$

The claim now follows from Lebesgue's differentiation theorem. $\square$

Remark 2.5.5. Just as in the non-weighted case, it is not possible to have a Sobolev type inequality in $W_w^{m,1}(\mathbf{R}^N)$ with any other exponent than the Sobolev exponent $1^* = N/(N-m)$. To see this, let $1 \le q < \infty$, and suppose that $w \in A_1$ and that there exists a constant C such that

$$\left(\int_{\mathbf{R}^N} |u|^q w \, dx\right)^{1/q} \le C \int_{\mathbf{R}^N} |\nabla^m u| w \, dx \qquad (2.5.9)$$

for every $u \in W_w^{m,1}(\mathbf{R}^N)$; we claim that $q = N/(N-m)$. If the constant C only depends on the A_1 constant of w, the claim follows by homogeneity: if $\delta > 0$, then

$$\left(\int_{\mathbf{R}^N} |u(x)|^q w(x) \, dx\right)^{1/q} = \delta^{N/q} \left(\int_{\mathbf{R}^N} |u(\delta x)|^q w(\delta x) \, dx\right)^{1/q}$$

$$\le C\delta^{N/q} \int_{\mathbf{R}^N} |\nabla^m(u(\delta x))| w(\delta x) \, dx$$

$$= C\delta^{N/q+m-N} \int_{\mathbf{R}^N} |\nabla^m u(x)| w(x) \, dx.$$

since the weights $w(x)$ and $w(\delta x)$, according to Remark 1.2.4.5, have the same A_1 constant. But this is only possible if $q = N/(N-m)$.

Now suppose that the constant in (2.5.9) has a more complicated dependence on w. By the same argument as in the proof of the corollary above, we get

$$w(B_r(x_0))^{1/q} \le C \frac{w(B_r(x_0))}{r^m} \qquad (2.5.10)$$

for every $x_0 \in \mathbf{R}^N$ and every $r > 0$. It follows that q has to be greater than 1. The inequality (2.5.10) may be rewritten as

$$\fint_{B_r(x_0)} w \, dx \ge C r^{mq'-N}.$$

By letting $r \to 0$ in the last inequality, we see that $mq' - N \ge 0$. If $r \ge 1$, we also have

$$C r^{mq'-N} \le \fint_{B_r(x_0)} w \, dx \le C w(B_1(x_0))$$

by the strong doubling property of w. This similarly implies that $mq' - N \le 0$. Thus $mq' - N = 0$, i.e., $q = N/(N-m)$.

2.6. Embeddings into $L_\mu^q(\Omega)$

The purpose of this section is to determine conditions on a measure μ so that

$$\int_{\mathbf{R}^N} |u| \, d\mu \le C \int_{\mathbf{R}^N} |\nabla^m u| w \, dx$$

or

$$\int_{\mathbf{R}^N} |u|\, d\mu \le C\|u\|_{W^{m,1}_w(\mathbf{R}^N)}$$

for every function $u \in C^\infty_0(\mathbf{R}^N)$, where $1 \le m < N$ and w is a given A_1 weight. In the first subsection, we give some background to this problem, while in the second, we present the actual results.

2.6.1. Introduction

V. G. Maz'ya proved in 1972 the following result (see [73, p. 209] and [76, pp. 54–55]). Let Ω be an open subset of $\mathbf{R}^N$, and let $1 \le q < \infty$. Suppose that μ is a positive Radon measure on Ω, which satisfies

$$M_1 = \sup \frac{\mu(\gamma)^{1/q}}{\mathcal{H}^{N-1}(\partial\gamma)} < \infty, \tag{2.6.1}$$

where the supremum is over all open, bounded subsets γ of Ω, with C^∞ boundaries, such that $\overline{\gamma} \subset \Omega$. Then the inequality

$$\left(\int_\Omega |u|^q\, d\mu\right)^{1/q} \le M_1 \int_\Omega |\nabla u|\, dx \tag{2.6.2}$$

holds for every $u \in C^\infty_0(\Omega)$. Moreover, M_1 is the best constant in (2.6.2).[†] This inequality was also proved independently by N. G. Meyers and W. P. Ziemer [81, pp. 1358–1359] in the context of BV functions and for $\Omega = \mathbf{R}^N$.

The inequality (2.6.2) was later extended by Maz'ya to several derivatives in the case $\Omega = \mathbf{R}^N$ (see [75, pp. 2001–2002] and [76, pp. 56–57]). Let $1 \le q < \infty$ and $1 \le m < \infty$, and suppose that

$$M_2 = \sup_{a \in \mathbf{R}^N,\ r>0} \frac{\mu(B_r(a))^{1/q}}{r^{N-m}} < \infty. \tag{2.6.3}$$

Then, for every $u \in C^\infty_0(\mathbf{R}^N)$,

$$\left(\int_{\mathbf{R}^N} |u|^q\, d\mu\right)^{1/q} \le C \int_{\mathbf{R}^N} |\nabla^m u|\, dx, \tag{2.6.4}$$

with $C = C'M_2$, where C' only depends on N. Conversely, if there exists a constant C such that the inequality (2.6.4) holds for every $u \in C^\infty_0(\mathbf{R}^N)$,

---

[†]For this to be perfectly true, we should change our definition of $\mathcal{H}^{N-1}$ slightly, so that $h^{N-1}(B_r(x)) = \nu_{N-1} r^{N-1}$, where ν_{N-1} is the $(N-1)$-dimensional measure of the unit ball in $\mathbf{R}^{N-1}$.

then $C \geq (N|B_1(0)|)^{-1} M_2$. Maz'ya also proved that the best constant in the inequality

$$\left(\int_{\mathbf{R}^N} |u|^q \, d\mu\right)^{1/q} \leq C \|u\|_{W^{m,1}(\mathbf{R}^N)} \tag{2.6.5}$$

for $u \in C_0^\infty(\mathbf{R}^N)$, in the same way is comparable to

$$\sup_{a \in \mathbf{R}^N, \, 0 < r \leq 1} \frac{\mu(B_r(a))^{1/q}}{r^{N-m}} < \infty$$

(see [76, pp. 59–60]).

We now mention some applications that illustrate the importance of this kind of inequalities. First, if μ is the restriction of the Lebesgue measure to an open set Ω and $q = N/(N-1)$, then the isoperimetric inequality implies that (2.6.1) holds with $M_1 = N^{-1}|B_1(0)|^{-1/N}$, so it follows that

$$\left(\int_\Omega |u|^{N/(N-1)} \, dx\right)^{(N-1)/N} \leq N^{-1}|B_1(0)|^{-1/N} \int_\Omega |\nabla u| \, dx$$

for $u \in C_0^\infty(\mathbf{R}^N)$, with the best constant in the right-hand side. This is the Sobolev type inequality for $\overset{\circ}{W}{}^{m,1}(\Omega)$. Another application is the following. Let Ω be bounded with smooth boundary, let $q = (N-1)/(N-m)$, where $1 \leq m < N$, and let μ be the $(N-1)$-dimensional Hausdorff measure on $\partial\Omega$. Then the supremum in (2.6.3) is finite, and thus

$$\left(\int_{\partial\Omega} |u|^{(N-1)/(N-m)} \, d\mathcal{H}^{N-1}\right)^{(N-m)/(N-1)} \leq C \int_{\mathbf{R}^N} |\nabla^m u| \, dx$$

for $u \in C_0^\infty(\mathbf{R}^N)$. This inequality can in turn be used to show that functions in $W^{m,1}(\Omega)$ have boundary values, i.e., traces, in $L^q(\partial\Omega)$. The inequalities (2.6.4) and (2.6.5) are also crucial for some applications of potential theory to the study of the space $W^{m,1}(\Omega)$.

Now a few words about weighted analogues of the inequality (2.6.2). Let Ω be open, $1 \leq q < \infty$, and let μ denote a positive Radon measure. Let w be a weight. It is easy to see that Maz'ya's proof of the equivalence between (2.6.1) and (2.6.2) can be used, without any changes, to show that

$$\left(\int_\Omega |u|^q \, d\mu\right)^{1/q} \leq M_3 \int_\Omega |\nabla u| w \, dx$$

for every $u \in C_0^\infty(\Omega)$ if and only if

$$M_3 = \sup \frac{\mu(\gamma)^{1/q}}{\int_{\partial\gamma} w \, d\mathcal{H}^{N-1}} < \infty,$$

where the family $\{\gamma\}$ is as above. Of course, some assumption about the regularity of w has to be made to ensure that the integrals of w over $\partial\gamma$ are defined for every γ; it suffices to assume that w be continuous. E. Stredulinsky has more generally shown that if ν is a positive Radon measure on Ω, then

$$\left(\int_\Omega |u|^q \, d\mu\right)^{1/q} \leq C \int_\Omega |\nabla u| \, d\nu \qquad (2.6.6)$$

for every $u \in C_0^\infty(\Omega)$ if and only if

$$M_4 = \sup \frac{\mu(\mathcal{L}_t)^{1/q}}{\operatorname{cap}_{1,1}^\nu(\mathcal{L}_t; \Omega)} < \infty, \qquad (2.6.7)$$

with the best constant in (2.6.6) comparable to M_4; see Stredulinsky [100, p. 34]. Here, the supremum is over all level sets $\mathcal{L}_t = \{x \; ; \; \varphi(x) \geq t\}$, $t \in \mathbf{R}$, for functions $\varphi \in C_0^\infty(\Omega)$ and the capacity $\operatorname{cap}_{1,1}^\nu(\mathcal{L}_t; \Omega)$ is defined by

$$\operatorname{cap}_{1,1}^\nu(\mathcal{L}_t; \Omega) = \inf \left\{ \int_\Omega |\nabla\psi| \, d\nu \; ; \; \psi \in C_0^\infty(\Omega), \; \psi(x) \geq 1 \text{ on } \mathcal{L}_t \right\}.$$

If ν is absolutely continuous with respect to the Lebesgue measure with density w, Stredulinsky [100, p. 36] has proved that the best constant in (2.6.6) is comparable to

$$\sup \frac{\mu(\gamma)^{1/q}}{\nu_{N-1}(\partial\gamma)}, \qquad (2.6.8)$$

where $\{\gamma\}$ is as before and the weighted surface measure $\nu_{N-1}(\partial\gamma)$ of $\partial\gamma$ is given by

$$\nu_{N-1}(\partial\gamma) = \liminf_{h \to 0, \, h > 0} \frac{1}{h} \int_{E_h'} w \, dx,$$

with $E_h' = \{x \notin \partial\gamma \; ; \; \operatorname{dist}(x, \partial\gamma) \leq h\}$. Thus, $\nu_{N-1}(\partial\gamma)$ is essentially the same as the weighted Minkowski content $\mathcal{M}_{w,*}^{N-1}(\partial\gamma)$ (cf. Definition 2.4.11).

Let μ be a positive Radon measure, $1 \leq q < \infty$, and w an A_1 weight. Consider the inequality

$$\left(\int_{\mathbf{R}^N} |u|^q \, d\mu\right)^{1/q} \leq C \int_{\mathbf{R}^N} |\nabla u| w \, dx \qquad (2.6.9)$$

for $u \in C_0^\infty(\mathbf{R}^N)$. We know that necessary and sufficient conditions for (2.6.9) to hold are given by (2.6.7) and (2.6.8) with $\Omega = \mathbf{R}^N$. Unfortunately, these conditions seem difficult to verify in our applications. In view of the equivalence between (2.6.3) and (2.6.4), we also expect that it should be possible to give a condition in terms of balls, either involving some capacity or some weighted surface measure (Hausdorff measure or lower Minkowski content) of balls. In Theorem 2.6.1 below, we give a necessary and sufficient condition for (2.6.9) in terms of balls and we will later, in Theorem 3.5.7, show that this condition is equivalent to a condition involving capacities of balls. We will also give necessary and sufficient conditions for weighted versions of the inequalities (2.6.4) and (2.6.5). Here, however, we have to assume that the exponent $q = 1$, the technical reason for this being the absence of a Sobolev type inequality for $W_w^{m,1}(\mathbf{R}^N)$.

2.6.2. Embedding theorems

The proofs presented below are adapted from the corresponding proofs in the monographs by V. G. Maz'ya [76] and W. P. Ziemer [108].

Theorem 2.6.1. *Let* $w \in A_1$, *and let* $1 \leq q < \infty$. *Suppose that* μ *is a positive Radon measure, satisfying*

$$M = \sup_{a \in \mathbf{R}^N,\, r>0} \frac{r\mu(B_r(a))^{1/q}}{w(B_r(a))} < \infty. \tag{2.6.10}$$

Then the inequality

$$\left(\int_{\mathbf{R}^N} |u|^q \, d\mu \right)^{1/q} \leq C \int_{\mathbf{R}^N} |\nabla u| w \, dx \tag{2.6.11}$$

holds for every $u \in C_0^\infty(\mathbf{R}^N)$ *with* $C = C'M$, *where* C' *only depends on* N, q, *and the* A_1 *constant of* w. *Conversely, if there exists a constant* C *such that* (2.6.11) *holds for every* $u \in C_0^\infty(\mathbf{R}^N)$, *then* $C \geq C'M$, *with* C' *as before. In particular,* M *is finite.*

Remark 2.6.2. It follows from Corollary 2.4.14 that

$$\mathcal{M}_{w,*}^{N-1}(\partial B_r(a)) \geq C \frac{w(B_r(a))}{r},$$

which implies that

$$\sup_{a \in \mathbf{R}^N,\, r>0} \frac{\mu(B_r(a))^{1/q}}{\mathcal{M}_{w,*}^{N-1}(\partial B_r(a))} \leq C \sup_{a \in \mathbf{R}^N,\, r>0} \frac{r\mu(B_r(a))^{1/q}}{w(B_r(a))}.$$

A necessary condition for (2.6.10) to hold for every $u \in C_0^\infty(\mathbf{R}^N)$ is thus

$$\sup_{a \in \mathbf{R}^N,\, r>0} \frac{\mu(B_r(a))^{1/q}}{\mathcal{M}_{w,*}^{N-1}(\partial B_r(a))} < \infty.$$

It is not known to the author whether this condition is sufficient or not.

Proof. We begin by showing that (2.6.10) implies (2.6.11). Let $u \in C_0^\infty(\mathbf{R}^N)$, and let $\mathcal{L}_t$ and $\mathcal{E}_t$, $t \geq 0$, as in the last section, denote the level set the level surface of $|u|$, respectively. According to the boxing inequality (Theorem 2.4.23), for a.e. $t \geq 0$, there is a countable covering $\{B_j\}$ of $\mathcal{L}_t$, satisfying

$$\sum_j \frac{w(B_j)}{r_j} \leq C \int_{\mathcal{E}_t} w \, d\mathcal{H}^{N-1},$$

where r_j is the radius of B_j. It follows that

$$\mu(\mathcal{L}_t)^{1/q} \leq \sum_j \mu(B_j)^{1/q} \leq M \sum_j \frac{w(B_j)}{r_j} \leq CM \int_{\mathcal{E}_t} w \, d\mathcal{H}^{N-1}. \qquad (2.6.12)$$

We now use the fact that if f is a nonnegative, decreasing function on $\mathbf{R}_+$, then

$$\int_0^\infty f(t)^p \, dt^p \leq \left(\int_0^\infty f(t) \, dt \right)^p \qquad (2.6.13)$$

for $1 \leq p < \infty$ (see Maz'ya [76, p. 49]). The inequality (2.6.11) then follows from (2.6.12) and (2.6.13) together with the co-area formula (Theorem 2.4.7):

$$\left(\int_{\mathbf{R}^N} |u|^q \, d\mu \right)^{1/q} = \left(\int_0^\infty \mu(\mathcal{L}_t) \, dt^q \right)^{1/q} \leq \int_0^\infty \mu(\mathcal{L}_t)^{1/q} \, dt$$

$$\leq CM \int_0^\infty \left(\int_{\mathcal{E}_t} w \, d\mathcal{H}^{N-1} \right) dt = CM \int_{\mathbf{R}^N} |\nabla u| w \, dx.$$

To finish, we will show that (2.6.10) is necessary for (2.6.11) to hold for every $u \in C_0^\infty(\mathbf{R}^N)$ and that the best constant in (2.6.11) is comparable to M. Let $\varphi \in C_0^\infty(\mathbf{R}^N)$ with support in $B_2(0)$ such that $\varphi = 1$ on $\overline{B_1(0)}$ and set $\psi(x) = \varphi((x - a)/r)$ for $x \in \mathbf{R}^N$, where $a \in \mathbf{R}^N$ and $r > 0$ are arbitrary but fixed. The inequality (2.6.11) and the doubling property of w implies that

$$\mu(B)^{1/q} \leq \left(\int_{\mathbf{R}^N} |\psi|^q \, d\mu \right)^{1/q} \leq C \int_{\mathbf{R}^N} |\nabla \psi| w \, dx$$

$$\leq CC' \frac{w(B_{2r}(a))}{r} \leq CC' \frac{w(B_r(a))}{r}.$$

Thus, $M \leq C'C$, which is exactly what we wanted to prove. $\square$

Theorem 2.6.3. *Let $w \in A_1$, and let m be an integer such that $1 \leq m < N$. Suppose that μ is a positive Radon measure, satisfying*

$$M = \sup_{a \in \mathbf{R}^N, \, r > 0} \frac{r^m \mu(B_r(a))}{w(B_r(a))} < \infty. \qquad (2.6.14)$$

Then the inequality

$$\int_{\mathbf{R}^N} |u| \, d\mu \leq C \int_{\mathbf{R}^N} |\nabla^m u| w \, dx \qquad (2.6.15)$$

holds for every $u \in C_0^\infty(\mathbf{R}^N)$ with $C = C'M$, where C' only depends on m, N, and the A_1 constant of w. Conversely, if there exists a constant C such that (2.6.15) holds for every $u \in C_0^\infty(\mathbf{R}^N)$, then $C \geq C'M$, with C' as before. In particular, M is finite.

Proof. The fact that (2.6.15) implies (2.6.14) is proved as in Theorem 2.6.1; we will show the converse using induction on m, the case $m = 1$ having been proved in Theorem 2.6.1. Suppose that (2.6.15) holds for all previous values of m, and let $1 < m < N$ and $u \in C_0^\infty(\mathbf{R}^N)$. Since for every $x \in \mathbf{R}^N$,

$$|u(x)| \leq \int_{\mathbf{R}^N} \frac{|\nabla u(y)|}{|x-y|^{N-1}} \, dy = C \mathcal{I}_1 |\nabla u(x)|$$

(see (2.2.3)), Fubini's theorem implies that

$$\int_{\mathbf{R}^N} |u| \, d\mu \leq C \int_{\mathbf{R}^N} |\nabla u| \mathcal{I}_1 \mu \, dy.$$

Set $d\nu(y) = \mathcal{I}_1 \mu(y) \, dy$. By the induction hypothesis, (2.6.15) follows if we can prove that

$$\sup_{a \in \mathbf{R}^N, \, r > 0} \frac{r^{m-1} \nu(B_r(a))}{w(B_r(a))} \leq CM. \qquad (2.6.16)$$

We have, for arbitrary $a \in \mathbf{R}^N$ and $r > 0$,

$$\nu(B_r(a)) = \int_{|y-a|<r} \left(\int_{|x-a|<2r} \frac{d\mu(x)}{|x-y|^{N-1}} \right) dy$$
$$+ \int_{|y-a|<r} \left(\int_{|x-a|\geq 2r} \frac{d\mu(x)}{|x-y|^{N-1}} \right) dy$$
$$= I_1 + I_2.$$

The first integral, I_1, is estimated just by changing the order of integration and using the fact that $B_r(a) \subset B_{3r}(x)$ if $|x-a| < 2r$:

$$I_1 = \int_{|x-a|<2r} \left(\int_{|y-a|<r} \frac{dy}{|x-y|^{N-1}} \right) d\mu(x)$$
$$\leq \int_{|x-a|<2r} \left(\int_{|x-y|<3r} \frac{dy}{|x-y|^{N-1}} \right) d\mu(x)$$
$$= Cr\mu(B_{2r}(a)). \qquad (2.6.17)$$

To estimate I_2, we note that $|x-y| \geq \frac{1}{2}|x-a|$, if $|y-a| < r$ and $|x-a| \geq 2r$. This implies that

$$I_2 \leq 2^{N-1} \int_{|y-a|<r} \left(\int_{|x-a|\geq 2r} \frac{d\mu(x)}{|x-a|^{N-1}} \right) dy$$
$$= Cr^N \int_{|x-a|\geq 2r} \frac{d\mu(x)}{|x-a|^{N-1}}.$$

With the aid of Lemma 2.4.3, we now rewrite the last integral as

$$(N-1)\int_{2r}^\infty \frac{\mu(B_t(a))}{t^{N-1}}\frac{dt}{t} - \frac{\mu(B_{2r}(a))}{(2r)^{N-1}}$$

and note that this difference is bounded by

$$(N-1)\int_{2r}^\infty \frac{t^m\mu(B_t(a))}{w(B_t(a))}\frac{w(B_t(a))}{t^N}\frac{1}{t^{m-1}}\frac{dt}{t}.$$

Hence, by using the assumption (2.6.14) and the strong doubling property of w, we obtain

$$I_2 \le CMr^N\frac{w(B_{2r}(a))}{(2r)^N}\int_{2r}^\infty \frac{1}{t^{m-1}}\frac{dt}{t} = CM\frac{w(B_{2r}(a))}{r^{m-1}}.$$

The estimates above now give (2.6.16):

$$\begin{aligned}
\frac{r^{m-1}\nu(B_r(a))}{w(B_r(a))} &= \frac{r^{m-1}}{w(B_r(a))}(I_1 + I_2)\\
&\le C\frac{r^{m-1}}{w(B_r(a))}\left(r\mu(B_{2r}(a)) + M\frac{w(B_{2r}(a))}{r^{m-1}}\right)\\
&\le C\frac{r^{m-1}}{w(B_{2r}(a))}\left(r\mu(B_{2r}(a)) + M\frac{w(B_{2r}(a))}{r^{m-1}}\right)\\
&\le CM. \quad \square
\end{aligned}$$

We will finally prove a weighted analogue of the inequality (2.6.5).

Theorem 2.6.4. *Let $w \in A_1$, and let m be an integer, $1 \le m < N$. Suppose that μ is a positive Radon measure, satisfying*

$$M = \sup_{a\in\mathbf{R}^N, 0<r\le 1} \frac{r^m\mu(B_r(a))}{w(B_r(a))} < \infty.$$

Then the inequality

$$\int_{\mathbf{R}^N} |u|\,d\mu \le C\|u\|_{W^{m,1}_w(\mathbf{R}^N)}, \tag{2.6.18}$$

holds for every $u \in C_0^\infty(\mathbf{R}^N)$ with $C = C'M$, where C' only depends on m, N, and the A_1 constant of w. Conversely, if there exists a constant C such that (2.6.18) holds for every $u \in C_0^\infty(\mathbf{R}^N)$, then $C \ge C'M$, with C' as before. In particular, M is finite.

The proof of Theorem 2.6.4 is based on the lemma below.

Lemma 2.6.5. *Let $w \in A_1$ and let m be an integer, $1 \leq m < N$. Suppose that μ is a positive Radon measure, supported in a ball B with radius R, satisfying*

$$M = \sup_{a \in B, \, 0 < r \leq R} \frac{r^m \mu(B_r(a))}{w(B_r(a))} < \infty. \qquad (2.6.19)$$

Then the inequality

$$\int_B |u| \, d\mu \leq C \int_B |\nabla^m u| w \, dx \qquad (2.6.20)$$

holds for every $u \in C_0^\infty(B)$, where C only depends on m, N, and the A_1 constant of w.

Remark 2.6.6. When $w = 1$, the above lemma is an easy corollary to Theorem 2.6.3. In fact, if we set

$$M_1 = \sup_{a \in \mathbf{R}^N, \, r > 0} \frac{\mu(B_r(a))}{r^{N-m}}, \quad M_2 = \sup_{a \in B, 0 < r \leq R} \frac{\mu(B_r(a))}{r^{N-m}},$$

then obviously $M_2 \leq M_1$. If $a \in \mathbf{R}^N$ and $r \geq R$ are arbitrary, then

$$\frac{\mu(B_r(a))}{r^{N-m}} \leq \frac{\mu(B)}{R^{N-m}} \leq M_2,$$

and if $a \notin B$ and $0 < r < R$, there exists an $x \in B$ such that $B \cap B_r(a) \subset B_r(x)$, so we have

$$\frac{\mu(B_r(a))}{r^{N-m}} \leq \frac{\mu(B_r(x))}{r^{N-m}} \leq M_2.$$

This shows that $M_1 \leq M_2$ and hence, $M_1 = M_2$.

The situation is more complicated for general A_1 weights. It is easily seen that if w is a suitable power weight and if $a \in B$, then $r^{-m} w(B_r(a))$ may tend to 0, as $r \to \infty$, which implies that the supremum in Theorem 2.6.3 is infinite for every nontrivial measure μ, although the supremum in (2.6.19) may be finite.

Proof of Lemma 2.6.5. In the case $m = 1$, the proof is identical to the proof of Theorem 2.6.1. We will prove the lemma for $1 < m < N$ using the same technique as in the proof of Theorem 2.6.3. Let $u \in C_0^\infty(B)$. If $x \in B$, then $|u(x)| \leq C \mathcal{I}_{1,2R} |\nabla u|(x)$, and hence

$$\int_B |u| \, d\mu \leq C \int_B |\nabla u| (\mathcal{I}_{1,2R}\mu) \, dy = C \int_B |\nabla u| \, d\nu.$$

The lemma follows if we can prove that

$$\sup_{a \in B, 0 < r \leq R} \frac{r^{m-1} \nu(B_r(a))}{w(B_r(a))} \leq CM.$$

We have, for $a \in B$ and $0 < r \le R$,

$$\nu(B_r(a)) = \int_{|y-a|<r} \left(\int_{|x-y|<2R} \frac{d\mu(x)}{|x-y|^{N-1}} \right) dy$$

$$\le \int_{|y-a|<r} \left(\int_{|x-a|<3R} \frac{d\mu(x)}{|x-y|^{N-1}} \right) dy$$

$$= \int_{|y-a|<r} \left(\int_{|x-a|<2r} \frac{d\mu(x)}{|x-y|^{N-1}} \right) dy$$

$$+ \int_{|y-a|<r} \left(\int_{2r \le |x-a|<3R} \frac{d\mu(x)}{|x-y|^{N-1}} \right) dy$$

$$= I_1 + I_2.$$

The first integral is estimated as before. Using (2.6.17) and the doubling property of w, we get that

$$\frac{r^{m-1} I_1}{w(B_r(a))} \le C \frac{r^m \mu(B_{2r}(a))}{w(B_r(a))} \le C \frac{(2r)^m \mu(B_{2r}(a))}{w(B_{2r}(a))}.$$

If $r < R/2$, the last quotient is majorized by CM. In the case $R/2 \le r \le R$, we use the strong doubling property of w to obtain

$$\frac{(2r)^m \mu(B_{2r}(a))}{w(B_{2r}(a))} \le C \frac{R^m \mu(B)}{w(B)} \le CM.$$

We now estimate the integral I_2:

$$I_2 \le Cr^N \int_{2r \le |x-a|<3R} \frac{d\mu(x)}{|x-a|^{N-1}}.$$

Moreover,

$$\int_{2r \le |x-a|<3R} \frac{d\mu(x)}{|x-a|^{N-1}} = (N-1) \int_{2r}^{3R} \frac{\mu(B_t(a))}{t^{N-1}} \frac{dt}{t}$$

$$+ \frac{\mu(B_{3R}(a))}{(3R)^{N-1}} - \frac{\mu(B_{2r}(a))}{(2r)^{N-1}}. \qquad (2.6.21)$$

To bound the second term in the right-hand side of (2.6.21), we use (2.6.19) and the strong doubling property of w as follows:

$$\frac{\mu(B_{3R}(a))}{(3R)^{N-1}} = \frac{\mu(B)}{(3R)^{N-1}} = 3^{1-N} \frac{R^m \mu(B)}{w(B)} \frac{w(B)}{R^N} \frac{1}{R^{m-1}}$$

$$\le CM \frac{w(B_r(a))}{r^N} \frac{1}{r^{m-1}} = CM r^{-N} \frac{w(B_r(a))}{r^{m-1}}. \qquad (2.6.22)$$

We will now show that the same bound is valid for the integral in the right hand side of (2.6.21). If $r < R/2$, then

$$\int_{2r}^{3R} \frac{\mu(B_t(a))}{t^{N-1}} \frac{dt}{t} = \int_{2r}^{R} \frac{\mu(B_t(a))}{t^{N-1}} \frac{dt}{t} + \int_{R}^{3R} \frac{\mu(B_t(a))}{t^{N-1}} \frac{dt}{t}.$$

As in the proof of Theorem 2.6.3, we have

$$\int_{2r}^{R} \frac{\mu(B_t(a))}{t^{N-1}} \frac{dt}{t} \le CMr^{-N} \frac{w(B_r(a))}{r^{m-1}}.$$

Also, by (2.6.22),

$$\int_{R}^{3R} \frac{\mu(B_t(a))}{t^{N-1}} \frac{dt}{t} \le C \frac{\mu(B)}{R^{N-1}} \le CMr^{-N} \frac{w(B_r(a))}{r^{m-1}}. \tag{2.6.23}$$

If $r \ge R/2$, then

$$\int_{2r}^{3R} \frac{\mu(B_t(a))}{t^{N-1}} \frac{dt}{t} \le \int_{R}^{3R} \frac{\mu(B_t(a))}{t^{N-1}} \frac{dt}{t},$$

so the desired inequality follows from (2.6.23). Putting all this together, we see that

$$I_2 \le CM \frac{w(B_r(a))}{r^{m-1}},$$

and the proof is completed. □

Proof of Theorem 2.6.4. We will prove that (2.6.18) holds whenever $M < \infty$. The proof in the converse direction is easy. Let $u \in C_0^\infty(\mathbf{R}^N)$. Let $\mathbf{R}^N = \bigcup_j B_j$, where B_j are open balls with radii $\frac{1}{2}$ such that $\{2B_j\}$ covers $\mathbf{R}^N$ with finite multiplicity, and let $\{\varphi_j\}$ be a partition of unity with respect to $\{2B_j\}$ such that $|D^\alpha \varphi_j| \le C$ for every j, $|\alpha| \le m$. Using Lemma 2.6.5, we then get

$$\int_{\mathbf{R}^N} |u| \, d\mu \le \sum_j \int_{2B_j} |\varphi_j u| \, d\mu \le CM \sum_j \int_{2B_j} |\nabla^m (\varphi_j u)| w \, dx$$

$$\le CM \sum_j \|u\|_{W_w^{m,1}(2B_j)} \le CM \|u\|_{W_w^{m,1}(\mathbf{R}^N)},$$

and we are done. □

Chapter 3

Potential theory

Weighted potential theory has its origin in the theory of degenerate elliptic partial differential equations. Such equations, where the degeneration is given by an A_p weight or, more generally, by a doubling weight, have been investigated by, among others, the American school, represented by E. B. Fabes, D. S. Jerison, C. E. Kenig, and R. P. Serapioni, and by the Finnish school, with names such as J. Heinonen, T. Kilpeläinen, and O. Martio. Important contributions to weighted potential theory have also been made by D. R. Adams, H. Aikawa, V. G. Maz'ya, E. Nieminen, and S. K. Vodop'yanov. With applications to weighted Sobolev spaces in mind, several results of potential theoretic nature are established in this chapter.[†]

The first section contains an inhomogeneous version of a norm inequality by B. Muckenhoupt and R. L. Wheeden. This inequality and some corollaries will be used for the investigation of weighted Bessel and Riesz capacities in later sections.

In Section 3.2, we summarize N. G. Meyers' theory of L^p capacities. A generalization of a theorem by Meyers on the existence of capacitary measures and potentials is also presented in this section.

The exceptional sets for functions in the space $W_w^{m,p}(\Omega)$ are measured by the Bessel capacity $B_{m,p}^w$, if $1 < p < \infty$, and by the Hausdorff capacity $\mathcal{H}_{w,1}^{N-m}$, if $p = 1$. These capacities are investigated in Section 3.3 and Section 3.4, respectively. In Section 3.5, it will be shown that the Bessel capacity and the Hausdorff capacity are equivalent to variational capacities that are associated with the norm in $W_w^{m,p}(\mathbf{R}^N)$.

In L. I. Hedberg's proof of the spectral synthesis theorem that was briefly discussed in the preface, the concept of "thinness" of a set plays a crucial role. In Section 3.6 and in Section 3.7, we investigate different definitions of thinness

---

[†]We will not here go into the long and interesting history of potential theory and, in particular, nonlinear potential theory. For this, the reader should consult, e.g., Adams–Hedberg [7], Maz'ya [76], or Ziemer [108].

and prove Kellogg type lemmas.

3.1. Norm inequalities for fractional integrals and maximal functions

The theorem below by B. Muckenhoupt and R. L. Wheeden [87, p. 267] was formulated in Section 1.2.7.

Theorem 3.1.1. *Suppose that $w \in A_\infty$, and let $0 < p < \infty$ and $0 < \alpha < N$. Then there exists a constant C, that only depends on N, p, and the A_∞ constants of w, such that*

$$\int_{\mathbf{R}^N} |\mathcal{I}_\alpha f|^p w \, dx \leq C \int_{\mathbf{R}^N} (M_\alpha f)^p w \, dx$$

for every measurable function f.

We intend here to show a corresponding inequality between the inhomogeneous Riesz potential and the inhomogeneous maximal function of a positive measure. This result will be the main ingredient in the proof of a Wolff type inequality, that we establish in Section 3.6.2. As a corollary, we obtain a similar inequality between the Bessel potential and the inhomogeneous maximal function. This inequality is then used in Section 3.3.1 to show that the Bessel capacity is equivalent to an inhomogeneous Riesz capacity.

Our main result reads as follows.

Theorem 3.1.2. *Let $w \in A_\infty$, and let $\alpha > 0$, $\rho > 0$, and $0 < p < \infty$. Then there exists a positive constant C, depending only on N, α, p, and the A_∞ constants of w, such that, for every measure $\mu \in \mathcal{M}^+(\mathbf{R}^N)$,*

$$\int_{\mathbf{R}^N} (\mathcal{I}_{\alpha,\rho}\mu)^p w \, dx \leq C \int_{\mathbf{R}^N} (M_{\alpha,\rho}\mu)^p w \, dx. \tag{3.1.1}$$

3.1.1. Proof of the main inequality and some corollaries

The proof we are about to present is a modification of a proof for the case $1 \leq p < \infty$ and $w = 1$ communicated to the author by D. R. Adams and L. I. Hedberg; see also [7]. We begin by proving a so called "good-λ inequality." Such inequalities were first used by D. L. Burkholder and R. F. Gundy [17].

Lemma 3.1.3. *Let $w \in A_\infty$, and let $\alpha > 0$, $\rho > 0$, and $0 < p < \infty$. Let μ be a positive measure. Then there exists a positive constant $a \geq 1$ for which, for every $\eta > 0$, there is an ε, $0 < \varepsilon \leq 1$, such that the inequality*

$$w(\{x \in \mathbf{R}^N \, ; \, \mathcal{I}_{\alpha,\rho}\mu(x) > a\lambda\}) \leq \eta w(\{x \in \mathbf{R}^N \, ; \, \mathcal{I}_{\alpha,\rho}\mu(x) > \lambda\})$$
$$+ w(\{x \in \mathbf{R}^N \, ; \, M_{\alpha,\rho}\mu(x) > \varepsilon\lambda\}) \tag{3.1.2}$$

holds for every $\lambda > 0$.

Proof. Let $E_\lambda = \{x \in \mathbf{R}^N \; ; \; \mathcal{I}_{\alpha,\rho}\mu(x) > \lambda\}$, $\lambda > 0$. Then E_λ is open because $\mathcal{I}_{\alpha,\rho}\mu$ is lower semicontinuous (this follows easily from Fatou's lemma). Using Whitney's theorem (see Stein [99, p. 16]), we may decompose E_λ into closed, dyadic cubes $\{Q_j\}$, with disjoint interiors, such that, for each cube Q_j,

$$\operatorname{diam} Q_j \leq \operatorname{dist}(Q_j, E_\lambda^c) \leq 4\operatorname{diam} Q_j.$$

We then subdivide the cubes with diameters $> \frac{1}{8}\rho$ into dyadic subcubes with diameters $\leq \frac{1}{8}\rho$ but $> \frac{1}{16}\rho$. Let $\{Q_j\}$ also denote the modified decomposition.

Suppose that we can show that there is a constant $a \geq 1$ such that, for every $\eta > 0$, there is an ε, $0 < \varepsilon \leq 1$, satisfying

$$w(\{x \in Q \; ; \; \mathcal{I}_{\alpha,\rho}\mu(x) > a\lambda, \; M_{\alpha,\rho}\mu(x) \leq \varepsilon\lambda\}) \leq \eta w(Q), \qquad (3.1.3)$$

for every $\lambda > 0$ and every $Q \in \{Q_j\}$ (where $\{Q_j\}$ is the decomposition of E_λ). This implies that

$$w(\{x \in Q \; ; \; \mathcal{I}_{\alpha,\rho}\mu(x) > a\lambda\}) \leq \eta w(Q) + w(\{x \in Q \; ; \; M_{\alpha,\rho}\mu(x) > \varepsilon\lambda\}),$$

and the lemma follows by summing over Q.

In order to prove (3.1.3), we note that, by the A_∞ condition, for every $\eta > 0$, there is a constant δ such that if Q is a cube and E is a measurable subset of Q with $|E| \leq \delta|Q|$, then $w(E) \leq \eta w(Q)$. It thus suffices to prove that, for a given $\eta > 0$, there exists an ε such that

$$|\{x \in Q \; ; \; \mathcal{I}_{\alpha,\rho}\mu(x) > a\lambda, \; M_{\alpha,\rho}\mu(x) \leq \varepsilon\lambda\}| \leq \delta|Q|, \qquad (3.1.4)$$

for every λ and every $Q \in \{Q_j\}$. Let $\eta > 0$ be arbitrary and let $Q \in \{Q_j\}$ be one of the undivided Whitney cubes. Let P be the ball concentric to Q with radius $6\operatorname{diam} Q$, and set $\mu_1 = \mu|_P$ and $\mu_2 = \mu - \mu_1$. For arbitrary a and ε, we define

$$M = \{x \in Q \; ; \; \mathcal{I}_{\alpha,\rho}\mu(x) > a\lambda, \; M_{\alpha,\rho}\mu(x) \leq \varepsilon\lambda\},$$
$$M_1 = \{x \in Q \; ; \; \mathcal{I}_{\alpha,\rho}\mu_1(x) > \tfrac{1}{2}a\lambda, \; M_{\alpha,\rho}\mu(x) \leq \varepsilon\lambda\},$$

and

$$M_2 = \{x \in Q \; ; \; \mathcal{I}_{\alpha,\rho}\mu_2(x) > \tfrac{1}{2}a\lambda, \; M_{\alpha,\rho}\mu(x) \leq \varepsilon\lambda\}.$$

Below, it will be shown that there is a constant C, depending only on N and α, so that

$$|M_1| \leq C\left(\frac{\varepsilon}{a}\right)^{N/(N-\alpha)}|Q| \qquad (3.1.5)$$

and, furthermore, that if $a \geq 4L^{N-\alpha}$, where L only depends on N, and if $\varepsilon \leq 1$, then $M_2 = \emptyset$. Fixing $a = \max\{4L^{N-\alpha}, 1\}$, we then find

$$|M| \leq |M_1| \leq C\left(\frac{\varepsilon}{a}\right)^{N/(N-\alpha)}|Q| \leq C\varepsilon^{N/(N-\alpha)}|Q|,$$

and this proves (3.1.4) if we choose $0 < \varepsilon \le \varepsilon_1 = \min\{(\delta/C)^{(N-\alpha)/N}, 1\}$. We will now show that (3.1.5) holds. We may assume that there is a point $x_0 \in Q$ such that $M_{\alpha,\rho}\mu(x_0) \le \varepsilon\lambda$. Then

$$|M_1| \le |\{x \in Q \,;\, \mathcal{I}_{\alpha,\rho}\mu_1(x) > \tfrac{1}{2}a\lambda\}| \le |\{x \in Q \,;\, \mathcal{I}_\alpha\mu_1(x) > \tfrac{1}{2}a\lambda\}|$$
$$\le C\left(\frac{2}{a\lambda}\int_{\mathbf{R}^N} d\mu_1\right)^{N/(N-\alpha)},$$

where the last inequality is the weak Sobolev inequality for Riesz potentials (see Stein [99, p. 120]). If we let $B = B_{8\,\mathrm{diam}\,Q}(x_0)$, then

$$\int_{\mathbf{R}^N} d\mu_1 = \int_P d\mu \le \int_B d\mu \le |B|^{(N-\alpha)/N} M_{\alpha,\rho}\mu(x_0) \le C|Q|^{(N-\alpha)/N}\varepsilon\lambda.$$

This proves (3.1.5).

Now, let x_1 be a point in E_λ^c such that $\mathrm{dist}(x_1, Q) \le 4\,\mathrm{diam}\,Q$. For $x \in Q$, we then have

$$\mathcal{I}_{\alpha,\rho}\mu_2(x) \le L^{N-\alpha}\int_{|x-y|<\rho}\frac{d\mu_2(y)}{|x_1-y|^{N-\alpha}}$$
$$\le L^{N-\alpha}\left(\int_{|x_1-y|<\rho}\frac{d\mu_2(y)}{|x_1-y|^{N-\alpha}} + \int_{|x-y|<\rho,\,|x_1-y|\ge\rho}\frac{d\mu_2(y)}{|x_1-y|^{N-\alpha}}\right)$$
$$\le L^{N-\alpha}\lambda + L^{N-\alpha}M_{\alpha,\rho}\mu(x),$$

where L only depends on N. Thus, if $M_{\alpha,\rho}\mu(x) \le \varepsilon\lambda$ and $0 < \varepsilon \le 1$, then $\mathcal{I}_{\alpha,\rho}\mu_2(x) \le 2L^{N-\alpha}\lambda$, so $M_2 = \varnothing$, if $a \ge 4L^{N-\alpha}$.

We now consider one of the cubes Q obtained by subdividing a Whitney cube. Then $\tfrac{1}{16}\rho < \mathrm{diam}\,Q \le \tfrac{1}{8}\rho$. Using the same notation as above, we have $|M_1| \le \rho|Q|$, if $0 < \varepsilon \le \varepsilon_1$. Furthermore, for $x \in Q$,

$$\mathcal{I}_{\alpha,\rho}\mu_2(x) \le \int_{11\rho/32\le|x-y|<\rho}\frac{d\mu(y)}{|x-y|^{N-\alpha}} \le \left(\frac{32}{11\rho}\right)^{N-\alpha}\int_{|x-y|<\rho}d\mu(y)$$
$$\le \left(\frac{32}{11}\right)^{N-\alpha}M_{\alpha,\rho}\mu(x).$$

If $x \in M_2$, we thus have

$$\mathcal{I}_{\alpha,\rho}\mu_2(x) \le \left(\frac{32}{11}\right)^{N-\alpha}\varepsilon\lambda.$$

This shows that $M_2 = \varnothing$, if $0 < \varepsilon \le \varepsilon_2 = \tfrac{1}{2}\left(\tfrac{11}{32}\right)^{N-\alpha}a$. Hence, if we fix $\varepsilon = \min\{\varepsilon_1, \varepsilon_2\}$, we obtain (3.1.4), and we are done. $\square$

We will now show how Theorem 3.1.2 follows from the lemma just proved.

Proof of Theorem 3.1.2. First assume that μ has compact support. After multiplying both sides of the inequality in the lemma by λ^{p-1} and integrating with respect to λ from 0 to $R > 0$, we find

$$\int_0^R \lambda^{p-1} w(\{x \; ; \mathcal{I}_{\alpha,\rho}\mu(x) > a\lambda\}) \, d\lambda \leq \eta \int_0^R \lambda^{p-1} w(\{x \; ; \mathcal{I}_{\alpha,\rho}\mu(x) > \lambda\}) \, d\lambda$$
$$+ \int_0^R \lambda^{p-1} w(\{x \; ; M_{\alpha,\rho}\mu(x) > \varepsilon\lambda\}) \, d\lambda,$$

and, after changing variables and using the fact that $a \geq 1$,

$$a^{-p} \int_0^{aR} \lambda^{p-1} \, w(\{x \; ; \mathcal{I}_{\alpha,\rho}\mu(x) > \lambda\}) \, d\lambda$$
$$\leq \eta \int_0^{aR} \lambda^{p-1} w(\{x \; ; \mathcal{I}_{\alpha,\rho}\mu(x) > \lambda\}) \, d\lambda$$
$$+ \varepsilon^{-p} \int_0^{\varepsilon R} \lambda^{p-1} w(\{x \; ; M_{\alpha,\rho}\mu(x) > \lambda\}) \, d\lambda.$$

Both integrals in the right-hand side of the last inequality are finite, since the fact that $\operatorname{supp}\mu$ is compact implies that both $\mathcal{I}_{\alpha,\rho}\mu$ and $M_{\alpha,\rho}\mu$ have compact support. Now choose $\eta = \frac{1}{2}a^{-p}$. Then

$$\frac{1}{2}a^{-p} \int_0^{aR} \lambda^{p-1} w(\{x \; ; \mathcal{I}_{\alpha,\rho}\mu(x) > \lambda\}) \, d\lambda$$
$$\leq \varepsilon^{-p} \int_0^{\varepsilon R} \lambda^{p-1} w(\{x \; ; M_{\alpha,\rho}\mu(x) > \lambda\}) \, d\lambda,$$

and inequality (3.1.1) in the statement of the theorem follows by letting $R \to \infty$.

If $\operatorname{supp}\mu$ is not compact, we let $\mu_m = \mu|_{B_m(0)}$ for $m = 1, 2, \dots$. Then $\mathcal{I}_{\alpha,\rho}\mu_m \nearrow \mathcal{I}_{\alpha,\rho}\mu$, as $m \to \infty$, so by monotone convergence and the first part of the proof,

$$\int_{\mathbf{R}^N} (\mathcal{I}_{\alpha,\rho}\mu)^p w \, dx = \lim_{m \to \infty} \int_{\mathbf{R}^N} (\mathcal{I}_{\alpha,\rho}\mu_m)^p w \, dx$$
$$\leq C \lim_{m \to \infty} \int_{\mathbf{R}^N} (M_{\alpha,\rho}\mu_m)^p w \, dx$$
$$\leq C \int_{\mathbf{R}^N} (M_{\alpha,\rho}\mu)^p w \, dx.$$

The proof is thus completed. $\square$

We next prove a slight amplification of the inequality (3.1.1).

Corollary 3.1.4. *Let $w \in A_\infty$, and let $\alpha > 0$, $\rho > 0$, and $0 < p < \infty$. Then there is a positive constant C, depending only on N, α, p, and the A_∞ constants of w, such that, for every measure $\mu \in \mathcal{M}^+(\mathbf{R}^N)$,*

$$\int_{\mathbf{R}^N} (\mathcal{I}_{\alpha,2\rho}\mu)^p w \, dx \leq C \int_{\mathbf{R}^N} (M_{\alpha,\rho}\mu)^p w \, dx. \tag{3.1.6}$$

Proof. Note that, for every $x \in \mathbf{R}^N$,

$$\mathcal{I}_{\alpha,2\rho}\mu(x) = \int_{|x-y|<\rho} \frac{d\mu(y)}{|x-y|^{N-\alpha}} + \int_{\rho \leq |x-y|<2\rho} \frac{d\mu(y)}{|x-y|^{N-\alpha}}$$

$$\leq \mathcal{I}_{\alpha,\rho}\mu(x) + \frac{\mu(B_{2\rho}(x))}{\rho^{N-\alpha}}.$$

If we can show that

$$\int_{\mathbf{R}^N} \left(\frac{\mu(B_{2\rho}(x))}{\rho^{N-\alpha}} \right)^p w(x) \, dx \leq C \int_{\mathbf{R}^N} (M_{\alpha,\rho}\mu)^p w \, dx, \tag{3.1.7}$$

the inequality (3.1.6) follows from the theorem. For this purpose, we subdivide $\mathbf{R}^N$ into closed cubes $\{Q_j\}$ such that each cube has diameter $\frac{1}{2}\rho$. Then

$$\int_{\mathbf{R}^N} \mu(B_{2\rho}(x))^p w(x) \, dx \leq \sum_j \int_{Q_j} \mu(B_{2\rho}(x))^p w(x) \, dx.$$

Suppose that $x \in Q_j$. Then $B_{2\rho}(x) \subset \bigcup_{k=1}^m Q_k^{(j)}$, where $Q_k^{(j)} \in \{Q_j\}$ and m only depends on N. Using this and the strong doubling property of w, we obtain

$$\sum_j \int_{Q_j} \mu(B_{2\rho}(x))^p w(x) \, dx \leq C \sum_j \sum_{k=1}^m \mu(Q_k^{(j)})^p w(Q_j)$$

$$\leq C \sum_j \sum_{k=1}^m \mu(Q_k^{(j)})^p w(Q_k^{(j)})$$

$$\leq C \sum_j \mu(Q_j)^p w(Q_j).$$

For every $x \in Q_j$, $Q_j \subset B_\rho(x)$, and thus

$$\frac{\mu(Q_j)}{\rho^{N-\alpha}} \leq \frac{\mu(B_\rho(x))}{\rho^{N-\alpha}} \leq M_{\alpha,\rho}\mu(x).$$

The estimates above now give the inequality (3.1.7):

$$\int_{\mathbf{R}^N} \left(\frac{\mu(B_{2\rho}(x))}{\rho^{N-\alpha}} \right)^p w(x) \, dx \leq C \sum_j \int_{Q_j} \left(\frac{\mu(Q_j)}{\rho^{N-\alpha}} \right)^p w(x) \, dx$$

$$\leq C \int_{\mathbf{R}^N} (M_{\alpha,\rho}\mu)^p w \, dx. \quad \square$$

3.1.2. An inequality for Bessel potentials

Corollary 3.1.5 below concerns Bessel potentials. We first summarize some properties of the Bessel kernel and Bessel potential, that will be used in the present and following sections; see Stein [99, pp. 130–133].

For $\alpha \in \mathbf{R}$, we define the *Bessel kernel* G_α as a tempered distribution through its Fourier transform:

$$\widehat{G}_\alpha(\xi) = (1 + 4\pi^2|\xi|^2)^{-\alpha/2}.$$

It is well-known that, for $\alpha > 0$, $G_\alpha \in L^1(\mathbf{R}^N)$ and that $G_\alpha(x)$ is positive, radial, decreasing, and continuous for $x \neq 0$. We will also use the following properties of the Bessel kernel: if $0 < \alpha < N$, then

$$G_\alpha(x) \leq CI_\alpha(x) \tag{3.1.8}$$

and

$$G_\alpha(x) = C|x|^{\alpha-N} + o(|x|^{\alpha-N}), \quad \text{as } |x| \to 0. \tag{3.1.9}$$

Furthermore, for $\alpha > 0$ and $|x| \geq 1$,

$$G_\alpha(x) \leq Ce^{-|x|/2}. \tag{3.1.10}$$

For $\alpha > 0$, the *Bessel potential* $\mathcal{G}_\alpha f$ of a measurable function f is defined by

$$\mathcal{G}_\alpha f = G_\alpha * f.$$

It is well-known that $\mathcal{G}_\alpha f \in \mathcal{S}$ if $f \in \mathcal{S}$. The Bessel potential $\mathcal{G}_\alpha \mu$ of a measure $\mu \in \mathcal{M}^+(\mathbf{R}^N)$ is similarly defined by

$$\mathcal{G}_\alpha \mu = G_\alpha * \mu.$$

Note that if $f \in L^p_w(\mathbf{R}^N)$, where $w \in A_p$, then $\mathcal{G}_\alpha f \in L^p_w(\mathbf{R}^N)$. Indeed, since $G_\alpha \in L^1(\mathbf{R}^N)$ and G_α is radial, we have $\mathcal{G}_\alpha f \leq CMf$, where $C = \int_{\mathbf{R}^N} G_\alpha \, dx$; see Stein [99, pp. 62–63]. The assertion thus follows from Muckenhoupt's maximal theorem.

Corollary 3.1.5. *Let $w \in A_\infty$, and let $\alpha > 0$, $\rho > 0$, and $0 < p < \infty$. Then there is a positive constant C, depending only on α, N, p, ρ, and the A_∞ constants of w, such that, for every measure $\mu \in \mathcal{M}^+(\mathbf{R}^N)$,*

$$\int_{\mathbf{R}^N} (\mathcal{G}_\alpha \mu)^p w \, dx \leq C \int_{\mathbf{R}^N} (M_{\alpha,\rho} \mu)^p w \, dx. \tag{3.1.11}$$

Proof. By the properties (3.1.8) and (3.1.10) of the Bessel kernel, we have

$$\mathcal{G}_\alpha \mu(x) \leq C\mathcal{I}_{\alpha,\rho} \mu(x) + C\rho^{\alpha-N} \int_{|x-y|\geq\rho} e^{-|x-y|/2} \, d\mu(y).$$

If $I(x)$ denotes the second integral in the right-hand side of this inequality, it suffices to show that

$$\rho^{(\alpha-N)p} \int_{\mathbf{R}^N} I(x)^p w(x)\, dx \le C \int_{\mathbf{R}^N} (M_{\alpha,\rho}\mu)^p w\, dx.$$

As in the proof of the previous corollary, we let $\{Q_j\}$ be a subdivision of $\mathbf{R}^N$ into closed cubes with diameters $\frac{1}{2}\rho$. Using Hölder's inequality, in the case $p > 1$, and the inequality

$$\left(\sum_j a_j\right)^p \le \sum_j a_j^p,$$

when $0 < p < 1$ (cf. (2.4.16)), we then obtain

$$I(x)^p \le \left(\sum_j e^{-\operatorname{dist}(x,Q_j)/2}\mu(Q_j)\right)^p \le C \sum_j e^{-C_p \operatorname{dist}(x,Q_j)/2}\mu(Q_j)^p,$$

where $C_p = 1$, if $p \ge 1$, and $C_p = p$, if $0 < p < 1$. Let x_j be the center of Q_j. For every $x \in \mathbf{R}^N$, $\operatorname{dist}(x,Q_j) \ge |x - x_j| - \frac{1}{4}\rho$, so we have

$$\int_{\mathbf{R}^N} e^{-C_p \operatorname{dist}(x,Q_j)/2}w(x)\, dx \le C \int_{\mathbf{R}^N} e^{-C_p|x-x_j|/2}w(x)\, dx.$$

Also,

$$\int_{\mathbf{R}^N} e^{-C_p|x-x_j|/2}w(x)\, dx = \sum_{k=0}^{\infty} \int_{k\rho \le |x-x_j| < (k+1)\rho} e^{-C_p|x-x_j|/2}w(x)\, dx$$

$$\le \sum_{k=0}^{\infty} e^{-C_p k\rho/2}w(B_{(k+1)\rho}(x_j)).$$

We know that $w \in A_q$ for some q, $1 < q < \infty$ (Theorem 1.2.10) . The strong doubling property of w then implies that

$$w(B_{(k+1)\rho}(x_j)) \le C(k+1)^{Nq}w(B_\rho(x_j)) \le C(k+1)^{Nq}w(Q_j).$$

It follows that

$$\int_{\mathbf{R}^N} e^{-C_p|x-x_j|/2}w(x)\, dx \le Cw(Q_j)\sum_{k=0}^{\infty} e^{-C_p k\rho/2}(k+1)^{Nq} = Cw(Q_j).$$

If now we use the estimates obtained so far, we find

$$\rho^{(\alpha-N)p}\int_{\mathbf{R}^N} I(x)^p w(x)\, dx \le C\sum_j \left(\int_{\mathbf{R}^N} e^{-C_p|x-x_j|/2}w(x)\, dx\right)\left(\frac{\mu(Q_j)}{\rho^{N-\alpha}}\right)^p$$

$$\le C\sum_j \left(\frac{\mu(Q_j)}{\rho^{N-\alpha}}\right)^p w(Q_j).$$

The rest of the proof is identical to that of Corollary 3.1.4. $\square$

3.2. Meyers' theory for L^p-capacities

In this section, we recall a few notions and results from N. G. Meyers' theory of L^p-capacities, as presented in his paper [77]; see also Chapter 2 in Adams–Hedberg [7]. We will also prove a generalization of an existence theorem by Meyers for capacitary measures and capacitary potentials.

3.2.1. Outline of Meyers' theory

Let ν be a fixed positive measure on $\mathbf{R}^M$. A *kernel* is a nonnegative, measurable function k on $\mathbf{R}^N \times \mathbf{R}^M$ such that the function $k(\cdot, y)$ is lower semicontinuous on $\mathbf{R}^N$ for ν-a.e. $y \in \mathbf{R}^M$ and such that the function $k(x, \cdot)$ is ν-measurable on $\mathbf{R}^M$ for every $x \in \mathbf{R}^N$.[†] In what follows, we shall adopt B. Fuglede's convention for denoting potentials [43, p. 149]. If μ is a positive Radon measure on $\mathbf{R}^N$, the *potential* $k(\mu, \cdot)$ of μ is defined by

$$k(\mu, y) = \int_{\mathbf{R}^N} k(x, y) \, d\mu(x), \quad \nu\text{-a.e. } y \in \mathbf{R}^M.$$

We also need to define potentials of signed Radon measures. Let $\mu \in \mathcal{M}(\mathbf{R}^N)$ with canonical Jordan decomposition $\mu = \mu^+ - \mu^-$. For ν-a.e. $y \in \mathbf{R}^M$, we then define

$$k(\mu, y) = k(\mu^+, y) - k(\mu^-, y),$$

if $k(\mu^+, y)$ or $k(\mu^-, y)$ is finite, and $k(\mu, y) = \infty$ otherwise. Let f be a nonnegative, ν-measurable function on $\mathbf{R}^M$. The potential $k(\cdot, f\nu)$ is then defined by

$$k(x, f\nu) = \int_{\mathbf{R}^M} k(x, y) f(y) \, d\nu(y), \quad x \in \mathbf{R}^N.$$

If $f = f^+ - f^-$ is an arbitrary, ν-measurable function on $\mathbf{R}^M$, we define

$$k(x, f\nu) = k(x, f^+\nu) - k(x, f^-\nu)$$

for $x \in \mathbf{R}^N$, whenever at least one of the terms in the right-hand side is finite.

For the remainder of this section, k will denote a fixed kernel.

Definition 3.2.1. Let $1 < p < \infty$. The L^p-*capacity* of a set $E \subset \mathbf{R}^N$ with respect to k and ν, $C_{k,\nu,p}(E)$, is the set function

$$C_{k,\nu,p}(E) = \inf\{\|f\|^p_{L^p_\nu(\mathbf{R}^M)} ; f \in L^p_\nu(\mathbf{R}^M)^+, k(x, f\nu) \geq 1 \text{ for every } x \in E\}.$$

Here, $L^p_\nu(\mathbf{R}^M)^+$ denotes the cone of nonnegative elements in $L^p_\nu(\mathbf{R}^M)$.

[†]Meyers requires that k should be lower semicontinuous on $\mathbf{R}^N \times \mathbf{R}^N$. The weaker assumption made here is, however, sufficient; see Adams–Hedberg [7].

An almost immediate consequence of this definition is the fact that if f belongs to $L_\nu^p(\mathbf{R}^M)$, then the potential $k(x, f\nu)$ is defined and finite for $C_{k,\nu,p}$-quasievery $x \in \mathbf{R}^N$, i.e., for $x \in \mathbf{R}^N \setminus E$, where $C_{k,\nu,p}(E) = 0$; see [77, p. 260].

Proposition 3.2.2. *If* $f \in L_\nu^p(\mathbf{R}^M)$ *and*

$$E = \{x \in \mathbf{R}^N \; ; \; k(x, |f|\nu) = \infty\},$$

then $C_{k,\nu,p}(E) = 0$.

Depending on the context, the abbreviation $C_{k,\nu,p}$-q.e. will be used for either $C_{k,\nu,p}$-quasievery or $C_{k,\nu,p}$-quasieverywhere.

The following *dual definition* of the capacity $C_{k,\nu,p}$ is a consequence of the minimax theorem; see [77, p. 273].

Proposition 3.2.3. *If* $E \subset \mathbf{R}^N$ *is a Suslin set,*[†] *then*

$$C_{k,\nu,p}(E)^{1/p} = \sup\{\mu(E) \; ; \; \mu \in \mathcal{M}^+(E), \; \|k(\mu, \cdot)\|_{L_\nu^{p'}(\mathbf{R}^M)} \leq 1\}. \qquad (3.2.1)$$

According to Proposition 3.2.4, $C_{k,\nu,p}$ is an *outer capacity*; see [77, p. 259].

Proposition 3.2.4. *For every set* $E \subset \mathbf{R}^N$,

$$C_{k,\nu,p}(E) = \inf\{C_{k,\nu,p}(G) \; ; \; G \supset E, \, G \text{ open}\}.$$

The capacity $C_{k,\nu,p}$ is also *subadditive*; see [77, p. 259].

Proposition 3.2.5. *If* E_n *is a subset of* $\mathbf{R}^N$ *for* $n = 1, 2, \ldots$, *then*

$$C_{k,\nu,p}\left(\bigcup_{n=1}^\infty E_n\right) \leq \sum_{n=1}^\infty C_{k,\nu,p}(E_n).$$

The next two propositions show that $C_{k,\nu,p}$ is continuous from the right on compact subsets of $\mathbf{R}^N$ [77, p. 263] and continuous from the left on arbitrary subsets of $\mathbf{R}^N$ [77, p. 262].

Proposition 3.2.6. *If* $K_1 \supset K_2 \supset \ldots$ *is a decreasing sequence of compact subsets of* $\mathbf{R}^N$, *then*

$$C_{k,\nu,p}\left(\bigcap_{n=1}^\infty K_n\right) = \lim_{n\to\infty} C_{k,\nu,p}(K_n).$$

Proposition 3.2.7. *If* $E_1 \subset E_2 \subset \ldots$ *is an increasing sequence of subsets of* $\mathbf{R}^N$, *then*

$$C_{k,\nu,p}\left(\bigcup_{n=1}^\infty E_n\right) = \lim_{n\to\infty} C_{k,\nu,p}(E_n).$$

[†]Suslin sets are sometimes also called *analytic*. The definition of these sets is rather long and cumbersome (see, e.g., Carleson [19, pp. 1–2]) and will therefore be omitted. We shall later use the fact that every Borel set is analytic.

The theorem below, a fundamental result in potential theory by G. Choquet [23], is used to prove a capacitability theorem for $C_{k,\nu,p}$.

Theorem 3.2.8. *Let C be an extended real-valued function, defined on the subsets of $\mathbf{R}^N$, such that*

(a) $C(\emptyset) = 0$;

(b) *if $E_1 \subset E_2$, then $C(E_1) \leq C(E_2)$;*

(c) *if $K_1 \supset K_2 \supset \ldots$ is a decreasing sequence of compact subsets of $\mathbf{R}^N$, then*

$$C(\cap_{n=1}^{\infty} K_n) = \lim_{n \to \infty} C(K_n);$$

(d) *if $E_1 \subset E_2 \subset \ldots$ is an increasing sequence of subsets of $\mathbf{R}^N$, then*

$$C(\cup_{n=1}^{\infty} E_n) = \lim_{n \to \infty} C(E_n).$$

Then every Suslin set and, in particular, every Borel set in $\mathbf{R}^N$ is capacitable. More precisely, if E is a Suslin set, then

$$C(E) = \sup\{C(K) \; ; \; K \subset E, \, K \, compact\} = \inf\{C(G) \; ; \; E \subset G, \, G \, open\}.$$

The properties (a) and (b) in Theorem 3.2.8 obviously hold for $C_{k,\nu,p}$, and it follows from Propositions 3.2.6 and 3.2.7 that (c) and (d) are also satisfied. We thus have the following *capacitability theorem*; see [77, p. 264].

Proposition 3.2.9. *If E is a Suslin set, then*

$$C_{k,\nu,p}(E) = \sup\{C_{k,\nu,p}(K) \; ; \; K \subset E, \, K \, compact\}$$
$$= \inf\{C_{k,\nu,p}(G) \; ; \; E \subset G, \, G \, open\}.$$

We next define the nonlinear potential and the nonlinear energy of a positive measure.

Definition 3.2.10. If $\mu \in \mathcal{M}^+(\mathbf{R}^N)$, then the *nonlinear potential* of μ, $V_{k,\nu,p}^{\mu}$, is the function

$$V_{k,\nu,p}^{\mu}(x) = k(x, k(\mu, \cdot)^{p'-1}\nu), \quad x \in \mathbf{R}^N.$$

Note that $V_{k,\nu,p}^{\mu}$, by Fatou's lemma, is lower semicontinuous.

Definition 3.2.11. If $\mu \in \mathcal{M}^+(\mathbf{R}^N)$, then the *nonlinear energy* of μ, $\mathcal{E}_{k,\nu,p}(\mu)$, is defined by

$$\mathcal{E}_{k,\nu,p}(\mu) = \int_{\mathbf{R}^N} V_{k,\nu,p}^{\mu} \, d\mu.$$

The set of measures in $\mathcal{M}^+(\mathbf{R}^N)$ with finite energy is denoted $\mathcal{E}_{k,\nu,p}^+(\mathbf{R}^N)$.

We will frequently use the fact that

$$\int_{\mathbf{R}^N} V_{k,\nu,p}^{\mu} \, d\mu = \|k(\mu, \, \cdot \,)\|_{L_\nu^{p'}(\mathbf{R}^M)}^{p'}, \tag{3.2.2}$$

for $\mu \in \mathcal{M}^+(\mathbf{R}^N)$. This identity is easily verified using Fubini's theorem. We let the quantity in the right-hand side of (3.2.2) define the energy of a signed measure.

Definition 3.2.12. If $\mu \in \mathcal{M}(\mathbf{R}^N)$, then the *nonlinear energy* of μ, $\mathcal{E}_{k,\nu,p}(\mu)$, is defined by

$$\mathcal{E}_{k,\nu,p}(\mu) = \|k(\mu, \, \cdot \,)\|_{L_\nu^{p'}(\mathbf{R}^M)}^{p'}.$$

The set of measures in $\mathcal{M}(\mathbf{R}^N)$ with finite energy is denoted $\mathcal{E}_{k,\nu,p}(\mathbf{R}^N)$.

Note that $\mathcal{E}_{k,\nu,p}(\, \cdot \,)$ is a seminorm on $\mathcal{E}_{k,\nu,p}(\mathbf{R}^N)$. Convergence with respect to this seminorm will be referred to as *strong convergence*, and fundamental sequences in $\mathcal{E}_{k,\nu,p}(\, \cdot \,)$ are, of course, called *Cauchy sequences*.

3.2.2. Capacitary measures and capacitary potentials

The existence of capacitary measures and the corresponding capacitary potentials will be crucial to our proof of Kellogg's lemma in Section 3.6. We first state a result concerning compact sets; see Meyers [77, p. 277].

Proposition 3.2.13. *If $K \subset \mathbf{R}^N$ is compact with $0 < C_{k,\nu,p}(K) < \infty$, then there is a measure $\mu^K \in \mathcal{M}^+(K)$, the capacitary measure of K, such that*

$$V_{k,\nu,p}^{\mu^K}(x) \geq 1 \quad \text{for } C_{k,\nu,p}\text{-q.e. } x \in K,$$
$$V_{k,\nu,p}^{\mu^K}(x) \leq 1 \quad \text{for every } x \in \operatorname{supp} \mu^K,$$

and

$$\mu^K(K) = C_{k,\nu,p}(K).$$

A consequence of this proposition is an equivalent formulation of $C_{k,\nu,p}$ on compact subsets of $\mathbf{R}^N$ due to V. G. Maz'ya and V. P. Havin [53].

Proposition 3.2.14. *If $K \subset \mathbf{R}^N$ is compact, then*

$$C_{k,\nu,p}(K) = \sup\{\mu(K) \, ; \, \mu \in \mathcal{M}^+(K), \, V_{k,\nu,p}^{\mu}(x) \leq 1 \text{ for every } x \in \operatorname{supp} \mu\}.$$

Proposition 3.2.13 can be extended to arbitrary sets if the kernel k and the measure ν satisfies some extra conditions; see Meyers [77, p. 264].

Proposition 3.2.15. *Suppose that the function $k(\,\cdot\,, \varphi\nu)$ is continuous on $\mathbf{R}^N$ and $\lim_{|x|\to\infty} k(x, \varphi\nu) = 0$ for every function φ in a dense subset of $L^p_\nu(\mathbf{R}^N)$. Let E be an arbitrary subset of $\mathbf{R}^N$ with $0 < C_{k,\nu,p}(E) < \infty$. Then there is a measure $\mu^E \in \mathcal{M}^+(\overline{E})$ such that*

$$V^{\mu^E}_{k,\nu,p}(x) \geq 1 \quad \text{for } C_{k,\nu,p}\text{-q.e. } x \in E,$$

$$V^{\mu^E}_{k,\nu,p}(x) \leq 1 \quad \text{for every } x \in \operatorname{supp}\mu^E,$$

and

$$\mu^E(\overline{E}) = C_{k,\nu,p}(E).$$

We will show below that a stronger result can be obtained by utilizing techniques that resemble the ones used in the classical case $p = 2$; see §III in Carleson [19] and, in particular, Chapter II in Landkof [69]. The assumptions in Proposition 3.2.15 will be replaced with the requirement that $\mathcal{E}^+_{k,\nu,p}(\mathbf{R}^N)$ be consistent.

Definition 3.2.16. We say that $\mathcal{E}^+_{k,\nu,p}(\mathbf{R}^N)$ is *consistent*, if every Cauchy sequence in $\mathcal{E}^+_{k,\nu,p}(\mathbf{R}^N)$ that is weakly convergent to some measure belonging to $\mathcal{E}^+_{k,\nu,p}(\mathbf{R}^N)$ also converges strongly to the same measure.[†]

Theorem 3.2.17. *Suppose that $\mathcal{E}^+_{k,\nu,p}(\mathbf{R}^N)$ is consistent. Then, for every set $E \subset \mathbf{R}^N$, there is a measure $\mu^E \in \mathcal{M}^+(\overline{E})$ such that*

$$V^{\mu^E}_{k,\nu,p}(x) \geq 1 \quad \text{for } C_{k,\nu,p}\text{-q.e. } x \in E,$$

$$V^{\mu^E}_{k,\nu,p}(x) \leq 1 \quad \text{for every } x \in \operatorname{supp}\mu^E,$$

and

$$\mu^E(\overline{E}) = C_{k,\nu,p}(E).$$

Remark 3.2.18. Suppose that $k(\,\cdot\,, \varphi\nu)$ is continuous on $\mathbf{R}^N$ for every function $\varphi \in D$, where D is a dense subset of $L^p_\nu(\mathbf{R}^N)$. We will show that $\mathcal{E}^+_{k,\nu,p}(\mathbf{R}^N)$ then is consistent. Let $\{\gamma_n\}^\infty_{n=1}$ be a Cauchy sequence in $\mathcal{E}^+_{k,\nu,p}(\mathbf{R}^N)$ such that γ_n converges weakly to $\gamma \in \mathcal{E}^+_{k,\nu,p}(\mathbf{R}^N)$. Since

$$\mathcal{E}_{k,\nu,p}(\gamma_m - \gamma_n) = \|k(\gamma_m, \cdot) - k(\gamma_n, \cdot)\|^{p'}_{L^{p'}_\nu(\mathbf{R}^M)},$$

there is a function $f \in L^{p'}_\nu(\mathbf{R}^N)$ such that $k(\gamma_n, \cdot) \to f$ in $L^{p'}_\nu(\mathbf{R}^N)$. We claim that $f = k(\gamma, \cdot)$. If φ is an arbitrary function in D, then

$$\int_{\mathbf{R}^M} f\varphi\,d\nu = \lim_{n\to\infty} \int_{\mathbf{R}^M} k(\gamma_n, \cdot)\varphi\,d\nu = \lim_{n\to\infty} \int_{\mathbf{R}^N} k(\,\cdot\,, \varphi\nu)\,d\gamma_n$$

[†] We have borrowed this designation from B. Fuglede [43, p. 167], who, in the linear case, terms a kernel k on $\mathbf{R}^N \times \mathbf{R}^N$ consistent if the property in the definition holds and, in addition, k is symmetric and the energy $k(\mu, \mu) = \int k(\mu, y)\,d\mu(y)$ is ≥ 0 for every $\mu \in \mathcal{M}(\mathbf{R}^N)$.

$$= \int_{\mathbf{R}^N} k(\,\cdot\,, \varphi\nu)\, d\gamma = \int_{\mathbf{R}^M} k(\gamma, \,\cdot\,) \varphi\, d\nu,$$

which proves the claim. Thus, Theorem 3.2.17 is indeed stronger than Proposition 3.2.15.

The proof of Theorem 3.2.17 is based on a proof by L. I. Hedberg [58] in a case not covered by Meyers' result.

We begin with some preliminary observations. Let us denote the dual capacity in the right-hand side of (3.2.1) by $c_{k,\nu,p}$. Thus, for Borel sets $E \subset \mathbf{R}^N$,

$$c_{k,\nu,p}(E) = \sup\{\mu(E) \; ; \; \mu \in \mathcal{M}^+(E), \; \|k(\mu, \,\cdot\,)\|_{L^{p'}_\nu(\mathbf{R}^M)} \le 1\}.$$

Notice that $c_{k,\nu,p}$ may equivalently be defined by

$$c_{k,\nu,p}(E)^{-p'} = \inf\{\mathcal{E}_{k,\nu,p}(\gamma) \; ; \; \gamma \in \mathcal{M}^+(E), \; \gamma(E) = 1\}; \qquad (3.2.3)$$

see [77, p. 272]. We will denote the right-hand side in this identity by $w_{k,\nu,p}(E)$. The following corollary is a reformulation of Proposition 3.2.13.

Corollary 3.2.19. *If $K \subset \mathbf{R}^N$ is compact with $0 < C_{k,\nu,p}(K) < \infty$, then there exists an extremal measure $\gamma^K \in \mathcal{M}^+(K)$ for the capacity $w_{k,\nu,p}(K)$, satisfying*

$$V^{\gamma^K}_{k,\nu,p}(x) \ge \mathcal{E}_{k,\nu,p}(\gamma^K) \quad \text{for } C_{k,\nu,p}\text{-q.e. } x \in K, \qquad (3.2.4)$$

$$V^{\gamma^K}_{k,\nu,p}(x) \le \mathcal{E}_{k,\nu,p}(\gamma^K) \quad \text{for every } x \in \operatorname{supp} \gamma^K, \qquad (3.2.5)$$

and

$$\mathcal{E}_{k,\nu,p}(\gamma^K) = C_{k,\nu,p}(K)^{-p'/p}. \qquad (3.2.6)$$

Proof. The measure μ^K in Proposition 3.2.13 is given by $\mu^K = C_{k,\nu,p}(K)^{1/p'}\mu$, where μ is an extremal measure for $c_{k,\nu,p}(K)$. If we let $\gamma^K = C_{k,\nu,p}(K)^{-1}\mu^K$, then γ^K is a probability measure since $\mu^K(K) = C_{k,\nu,p}(K)$, and

$$\mathcal{E}_{k,\nu,p}(\gamma^K) = C_{k,\nu,p}(K)^{-p'}\mathcal{E}_{k,\nu,p}(\mu^K) = C_{k,\nu,p}(K)^{-p'+1}\mathcal{E}_{k,\nu,p}(\mu)$$
$$= C_{k,\nu,p}(K)^{-p'/p}. \qquad (3.2.7)$$

The last identity together with Proposition 3.2.3 and the formula (3.2.3) imply that γ^K is an extremal measure for $w_{k,\nu,p}(K)$. It also follows from (3.2.7) that

$$V^{\gamma^K}_{k,\nu,p}(x) = C_{k,\nu,p}(K)^{1-p'}V^{\mu^K}_{k,\nu,p}(x) = \mathcal{E}_{k,\nu,p}(\gamma^K)\, V^{\mu^K}_{k,\nu,p}(x).$$

The properties (3.2.4) and (3.2.5) are thus consequences of the corresponding properties for $V^{\mu^K}_{k,\nu,p}$. $\square$

The inequalities below are due to J. A. Clarkson [27] (see also, e.g., Hewitt–Stromberg [60, pp. 225–227]), who used them to establish the uniform convexity of $L^p(\Omega)$ for $1 < p < \infty$.

Lemma 3.2.20. *Suppose that f and g belong to $L_\nu^p(\Omega)$, where $1 < p < \infty$ and Ω is an open subset of $\mathbf{R}^M$.*

(a) *If $1 < p < 2$, then*

$$\left\|\frac{f-g}{2}\right\|_{L_\nu^p(\Omega)}^{p'} + \left\|\frac{f+g}{2}\right\|_{L_\nu^p(\Omega)}^{p'} \le \left(\frac{1}{2}\|f\|_{L_\nu^p(\Omega)}^p + \frac{1}{2}\|g\|_{L_\nu^p(\Omega)}^p\right)^{p'-1}.$$

(b) *If $2 \le p < \infty$, then*

$$\left\|\frac{f-g}{2}\right\|_{L_\nu^p(\Omega)}^p + \left\|\frac{f+g}{2}\right\|_{L_\nu^p(\Omega)}^p \le \frac{1}{2}\|f\|_{L_\nu^p(\Omega)}^p + \frac{1}{2}\|g\|_{L_\nu^p(\Omega)}^p.$$

Since the energy $\mathcal{E}_{k,\nu,p}(\gamma)$ of a measure $\gamma \in \mathcal{M}(\mathbf{R}^N)$ by definition is

$$\mathcal{E}_{k,\nu,p}(\gamma) = \|k(\gamma,\,\cdot\,)\|_{L_\nu^{p'}(\mathbf{R}^M)}^{p'},$$

Lemma 3.2.20 yields the following inequalities that for $p \ne 2$ will serve as substitutes for the parallelogram identity in $L_\nu^2(\mathbf{R}^N)$.

Corollary 3.2.21. *Suppose that γ_1 and γ_2 belong to $\mathcal{M}(\mathbf{R}^N)$.*

(a) *If $1 < p \le 2$, then*

$$\mathcal{E}_{k,\nu,p}\left(\frac{\gamma_1 - \gamma_2}{2}\right) + \mathcal{E}_{k,\nu,p}\left(\frac{\gamma_1 + \gamma_2}{2}\right) \le \frac{1}{2}\mathcal{E}_{k,\nu,p}(\gamma_1) + \frac{1}{2}\mathcal{E}_{k,\nu,p}(\gamma_2).$$

(b) *If $2 < p < \infty$, then*

$$\mathcal{E}_{k,\nu,p}\left(\frac{\gamma_1 - \gamma_2}{2}\right)^{p/p'} + \mathcal{E}_{k,\nu,p}\left(\frac{\gamma_1 + \gamma_2}{2}\right)^{p/p'}$$
$$\le \left(\frac{1}{2}\mathcal{E}_{k,\nu,p}(\gamma_1) + \frac{1}{2}\mathcal{E}_{k,\nu,p}(\gamma_2)\right)^{p-1}.$$

We next show that a positive measure with finite energy cannot lay mass on a compact set with capacity 0.

Lemma 3.2.22. *Suppose that $K \subset \mathbf{R}^N$ is compact with $C_{k,\nu,p}(K) = 0$. Then $\mu(K) = 0$ for any $\mu \in \mathcal{E}_{k,\nu,p}^+(\mathbf{R}^N)$.*

Proof. If $\mu \in \mathcal{E}_{k,\nu,p}^+(\mathbf{R}^N)$ and $\mu(K) > 0$, then the measure $\gamma = \mu(K)^{-1}\mu|_K$ belongs to $\mathcal{M}^+(K) \cap \mathcal{E}_{k,\nu,p}^+(\mathbf{R}^N)$ and has mass 1. It then follows from Proposition 3.2.3 and the identity (3.2.3) that

$$C_{k,\nu,p}(K)^{p'/p} = w_{k,\nu,p}(K)^{-1} \ge \mathcal{E}_{k,\nu,p}(\gamma)^{-1} > 0,$$

which is a contradiction. $\square$

We will also need the fact that the energy functional $\mathcal{E}_{k,\nu,p}(\,\cdot\,)$ is lower semicontinuous with respect to weak convergence.

Lemma 3.2.23. *Suppose that $\{\gamma_n\}_{n=1}^{\infty}$ is a sequence of measures in $\mathcal{M}^+(\mathbf{R}^N)$, that converges weakly to $\gamma \in \mathcal{M}^+(\mathbf{R}^N)$. Then*

$$\liminf_{n \to \infty} \mathcal{E}_{k,\nu,p}(\gamma_n) \geq \mathcal{E}_{k,\nu,p}(\gamma). \tag{3.2.8}$$

Proof. To prove the inequality (3.2.8), one first shows that

$$\liminf_{n \to \infty} k(\gamma_n, y) \geq k(\gamma, y)$$

for ν-a.e. $y \in \mathbf{R}^M$. This is a consequence of the lower semicontinuity of the function $\mathbf{R}^N \ni x \mapsto k(x,y)$; see Landkof [69, p. 8]. The assertion now follows from Fatou's lemma:

$$\mathcal{E}_{k,\nu,p}(\gamma) = \int_{\mathbf{R}^M} k(\gamma, y)^{p'} \, d\nu(y) \leq \liminf_{n \to \infty} \int_{\mathbf{R}^M} k(\gamma_n, y)^{p'} \, d\nu(y)$$
$$= \liminf_{n \to \infty} \mathcal{E}_{k,\nu,p}(\gamma_n). \quad \square$$

For the lemma below, a potential theoretic equivalent to Egorov's theorem in measure theory; see Meyers [77, p. 261].

Lemma 3.2.24. *Suppose that the sequence $\{f_n\}_{n=1}^{\infty}$ converges to f in $L_\nu^p(\mathbf{R}^M)$. Then there is a subsequence $\{f_{n_j}\}_{j=1}^{\infty}$ such that*

$$k(x, f_{n_j}\nu) \to k(x, f\nu), \quad \text{as } j \to \infty,$$

for $C_{k,\nu,p}$-q.e. $x \in \mathbf{R}^N$, uniformly outside an open set of arbitrarily small $C_{k,\nu,p}$-capacity.

With the aid of this lemma, we now show that strong convergence of a sequence of measures in $\mathcal{E}_{k,\nu,p}^+(\mathbf{R}^N)$ implies convergence $C_{k,\nu,p}$-q.e. of a subsequence of the corresponding nonlinear potentials. The proof follows closely a similar proof by V. G. Maz'ya and V. P. Havin [53, p. 104].

Lemma 3.2.25. *Suppose that $\{\gamma_n\}_{n=1}^{\infty}$ is a sequence of measures in $\mathcal{E}_{k,\nu,p}^+(\mathbf{R}^N)$, that converges strongly to $\gamma \in \mathcal{E}_{k,\nu,p}^+(\mathbf{R}^N)$. Then there is a subsequence $\{\gamma_{n_j}\}_{j=1}^{\infty}$ such that*

$$V_{k,\nu,p}^{\gamma_{n_j}}(x) \to V_{k,\nu,p}^{\gamma}(x), \quad \text{as } j \to \infty,$$

for $C_{k,\nu,p}$-q.e. $x \in \mathbf{R}^N$.

Proof. We will use the following elementary inequalities. Let A and B be two nonnegative real numbers. If $1 < p < 2$, then

$$|A^{p'-1} - B^{p'-1}| \leq \frac{1}{p-1} \max\{A, B\}^{(2-p)/(p-1)}|A - B|,$$

and if $2 \leq p < \infty$, then

$$|A^{p'-1} - B^{p'-1}| \leq |A - B|^{p'-1}.$$

Let us first consider the simplest case $2 \leq p < \infty$. An application of the last inequality shows that

$$\int_{\mathbf{R}^M} |k(\gamma, y)^{p'-1} - k(\gamma_n, y)^{p'-1}|^p \, d\nu(y) \leq \int_{\mathbf{R}^M} |k(\gamma, y) - k(\gamma_n, y)|^{p'} \, d\nu(y)$$
$$= \mathcal{E}_{k,\nu,p}(\gamma - \gamma_n).$$

Thus, $k(\gamma_n, \cdot)^{p'-1}$ converges to $k(\gamma, \cdot)^{p'-1}$ in $L^p_\nu(\mathbf{R}^M)$, and the conclusion follows from Lemma 3.2.24.

When $1 < p < 2$, we obtain in the same way that

$$\int_{\mathbf{R}^M} |k(\gamma, y)^{p'-1} - k(\gamma_n, y)^{p'-1}|^p \, d\nu(y)$$
$$\leq \frac{1}{(p-1)^p} \int_{\mathbf{R}^M} \max\{k(\gamma, y)^{p'}, k(\gamma_n, y)^{p'}\}^{2-p} |k(\gamma, y) - k(\gamma_n, y)|^p \, d\nu(y)$$
$$\leq \frac{1}{(p-1)^p} \left(\int_{\mathbf{R}^M} (k(\gamma, y)^{p'} + k(\gamma_n, y)^{p'}) \, d\nu(y) \right)^{2-p}$$
$$\times \left(\int_{\mathbf{R}^M} |k(\gamma, y) - k(\gamma_n, y)|^{p'} \, d\nu(y) \right)^{p-1}$$
$$= \frac{1}{(p-1)^p} (\mathcal{E}_{k,\nu,p}(\gamma) + \mathcal{E}_{k,\nu,p}(\gamma_n))^{2-p} \mathcal{E}_{k,\nu,p}(\gamma - \gamma_n)^{p-1}.$$

But, by Minkowski's inequality,

$$\mathcal{E}_{k,\nu,p}(\gamma_n)^{1/p'} \leq \mathcal{E}_{k,\nu,p}(\gamma - \gamma_n)^{1/p'} + \mathcal{E}_{k,\nu,p}(\gamma)^{1/p'} \leq 2 \mathcal{E}_{k,\nu,p}(\gamma)^{1/p'},$$

if n is large enough. This implies that

$$\int_{\mathbf{R}^M} |k(\gamma, y)^{p'-1} - k(\gamma_n, y)^{p'-1}|^p \, d\nu(y) \leq C \mathcal{E}_{k,\nu,p}(\gamma)^{2-p} \mathcal{E}_{k,\nu,p}(\gamma - \gamma_n)^{p-1}$$

for large n, and the proof is again completed by appealing to Lemma 3.2.24. $\square$

The next proposition extends Corollary 3.2.19 to general sets. Theorem 3.2.17 follows immediately from this proposition. Indeed, the measure μ^E is given by $\mu^E = C_{k,\nu,p}(E) \gamma^E$, where γ^E is the measure in the proposition.

Proposition 3.2.26. *Suppose that $\mathcal{E}_{k,\nu,p}(\mathbf{R}^N)$ is consistent. Then, for every subset E of $\mathbf{R}^N$ with $0 < C_{k,\nu,p}(E) < \infty$, there exists a measure $\gamma^E \in \mathcal{M}^+(\overline{E})$ such that*

$$V_{k,\nu,p}^{\gamma^E}(x) \geq \mathcal{E}_{k,\nu,p}(\gamma^E) \quad \text{for } C_{k,\nu,p}\text{-q.e. } x \in E, \tag{3.2.9}$$

$$V_{k,\nu,p}^{\gamma^E}(x) \leq \mathcal{E}_{k,\nu,p}(\gamma^E) \quad \text{for every } x \in \text{supp} \, \gamma^E, \tag{3.2.10}$$

and

$$\mathcal{E}_{k,\nu,p}(\gamma^E) = C_{k,\nu,p}(E)^{-p'/p}. \tag{3.2.11}$$

Proof. We first assume that E is open. Let $\{K_n\}_{n=1}^{\infty}$ be an increasing sequence of compact subsets of E such that $E = \bigcup_{n=1}^{\infty} K_n$ and $C_{k,\nu,p}(K_n)$ tends to $C_{k,\nu,p}(E)$, as $n \to \infty$. If $\gamma_n = \gamma^{K_n}$ are the corresponding capacitary measures, then since each γ_n is a probability measure, there is a subsequence of $\{\gamma_n\}_{n=1}^{\infty}$, which we still denote by $\{\gamma_n\}_{n=1}^{\infty}$, such that γ_n converges weakly to a measure $\gamma^E \in \mathcal{M}^+(\overline{E})$.[†] Lemma 3.2.23 then shows that

$$\liminf_{n \to \infty} \mathcal{E}_{k,\nu,p}(\gamma_n) \geq \mathcal{E}_{k,\nu,p}(\gamma^E).$$

But

$$\mathcal{E}_{k,\nu,p}(\gamma_n) = C_{k,\nu,p}(K_n)^{-p'/p} \to C_{k,\nu,p}(E)^{-p'/p},$$

as $n \to \infty$, so we have

$$\mathcal{E}_{k,\nu,p}(\gamma^E) \leq C_{k,\nu,p}(E)^{-p'/p} = w_{k,\nu,p}(E).$$

If we can show that $\gamma^E(\overline{E}) = 1$, it thus follows that γ^E is an extremal measure for $w_{k,\nu,p}(E)$.

We will next show that γ_n tends to γ^E strongly, and first claim that the sequence $\{\gamma_n\}_{n=1}^{\infty}$ is Cauchy. If $m < n$, then the probability measure $\frac{1}{2}(\gamma_m + \gamma_n)$ belongs to $\mathcal{M}^+(K_n)$, so by the definition of γ_n,

$$\mathcal{E}_{k,\nu,p}\left(\frac{\gamma_m + \gamma_n}{2}\right) \geq \mathcal{E}_{k,\nu,p}(\gamma_n).$$

First suppose that $1 < p \leq 2$. Then according to the first inequality in Corollary 3.2.21,

$$\begin{aligned}
\mathcal{E}_{k,\nu,p}\left(\frac{\gamma_m - \gamma_n}{2}\right) &\leq \frac{1}{2}\mathcal{E}_{k,\nu,p}(\gamma_m) + \frac{1}{2}\mathcal{E}_{k,\nu,p}(\gamma_n) - \mathcal{E}_{k,\nu,p}\left(\frac{\gamma_m + \gamma_n}{2}\right) \\
&\leq \frac{1}{2}\mathcal{E}_{k,\nu,p}(\gamma_m) - \frac{1}{2}\mathcal{E}_{k,\nu,p}(\gamma_n) \\
&= \frac{1}{2}C_{k,\nu,p}(K_m)^{-p'/p} - \frac{1}{2}C_{k,\nu,p}(K_n)^{-p'/p},
\end{aligned}$$

which proves the claim for $1 < p \leq 2$. The argument is almost identical in the remaining case, so we omit the details. By the assumption on the consistency of $\mathcal{E}_{k,\nu,p}^+(\mathbf{R}^N)$, we may conclude that γ_n tends to γ^E strongly. Note that it follows that $\lim_{n \to \infty} \mathcal{E}_{k,\nu,p}(\gamma_n) = \mathcal{E}_{k,\nu,p}(\gamma^E)$.

Lemma 3.2.25 now implies the existence of a subsequence $\{V_{k,\nu,p}^{\gamma_{n_j}}\}_{j=1}^{\infty}$ such that $V_{k,\nu,p}^{\gamma_{n_j}}$ converges $C_{k,\nu,p}$-q.e. to $V_{k,\nu,p}^{\gamma^E}$. Let A be the exceptional set. We

---

[†]Here we use the well-known fact that if μ_n converges weakly to μ, then

$$\operatorname{supp}\mu = \bigcap_{m=1}^{\infty} \overline{\bigcup_{n=m}^{\infty} \operatorname{supp}\mu_n};$$

see Landkof [69, p. 9].

know from Corollary 3.2.19 that

$$V_{k,\nu,p}^{\gamma_{n_j}}(x) \geq \mathcal{E}_{k,\nu,p}(\gamma_{n_j})$$

for $x \in K_{n_j} \setminus B_{n_j}$, where $C_{k,\nu,p}(B_{n_j}) = 0$. If we define $B = \bigcup_{j=1}^{\infty} B_{n_j}$, then by Proposition 3.2.5, $C_{k,\nu,p}(B) = 0$, and for $x \in E \setminus (A \cup B)$,

$$V_{k,\nu,p}^{\gamma^E}(x) = \lim_{j \to \infty} V_{k,\nu,p}^{\gamma_{n_j}}(x) \geq \lim_{j \to \infty} \mathcal{E}_{k,\nu,p}(\gamma_{n_j}) = \mathcal{E}_{k,\nu,p}(\gamma^E).$$

This proves the inequality (3.2.9).

We now verify (3.2.10), and begin by showing that, for every $x \in \operatorname{supp} \gamma^E$, there exists a sequence $\{x_m\}_{m=1}^{\infty}$, that converges to x, consisting of points x_m belonging to $\bigcup_{n=1}^{\infty} \operatorname{supp} \gamma_n \setminus A$. Since $\operatorname{supp} \gamma^E = \overline{\bigcup_{n=1}^{\infty} \operatorname{supp} \gamma_n}$, we can find points $x_m \in \bigcup_{n=1}^{\infty} \operatorname{supp} \gamma_n$ such that $x_m \to x$, as $m \to \infty$. Suppose that some $x_m \in \operatorname{supp} \gamma_{n_m}$ belongs to A. Then arbitrarily close to each x_m there has to be points in $\operatorname{supp} \gamma_{n_m} \setminus A$. For otherwise, the compact set $\overline{B_r(x_m)} \cap \operatorname{supp} \gamma_{n_m}$ would be a subset of A for some $r > 0$ and consequently have capacity 0 which, according to Lemma 3.2.22, contradicts the assumption that $x_m \in \operatorname{supp} \gamma_{n_m}$. We may thus replace each $x_m \in \operatorname{supp} \gamma_{n_m} \cap A$ with a new point, still denoted by x_m, such that $x_m \in \operatorname{supp} \gamma_{n_m} \setminus A$ and so that the modified sequence $\{x_m\}_{m=1}^{\infty}$ converges to x. Using this, we obtain

$$V_{k,\nu,p}^{\gamma^E}(x) \leq \liminf_{m \to \infty} V_{k,\nu,p}^{\gamma^E}(x_m) = \liminf_{m \to \infty} \lim_{j \to \infty} V_{k,\nu,p}^{\gamma_{n_j}}(x_m)$$
$$\leq \lim_{j \to \infty} \mathcal{E}_{k,\nu,p}(\gamma_{n_j}) = \mathcal{E}_{k,\nu,p}(\gamma^E).$$

It remains to prove that $\gamma^E(\overline{E}) = 1$. Let $K \subset \overline{E}$ be compact, and let $\varphi \in C_0(\mathbf{R}^N)$ be such that $0 \leq \varphi \leq 1$ and $\varphi = 1$ in a neighbourhood of K. Then

$$\gamma^E(K) \leq \int_{\overline{E}} \varphi \, d\gamma^E = \lim_{n \to \infty} \int_{\overline{E}} \varphi \, d\gamma_n \leq 1,$$

which, by the inner regularity of γ^E, implies that $\gamma^E(\overline{E}) \leq 1$. The reverse inequality follows from (3.2.10). We have

$$\mathcal{E}_{k,\nu,p}(\gamma^E) = \int_{\overline{E}} V_{k,\nu,p}^{\gamma^E} \, d\gamma^E \leq \mathcal{E}_{k,\nu,p}(\gamma^E) \, \gamma^E(\overline{E}),$$

and hence, $\gamma^E(\overline{E}) \geq 1$.

We now turn to the case of general sets E. Let $\{G_n\}_{n=1}^{\infty}$ be a decreasing sequence of open sets, satisfying $G_n \supset E$ and $\overline{E} = \bigcap_{n=1}^{\infty} \overline{G_n}$, and such that $C_{k,\nu,p}(G_n) \to C_{k,\nu,p}(E)$, as $n \to \infty$. If $\gamma_n \in \mathcal{M}^+(\overline{G_n})$ denotes the capacitary measure for G_n, one shows as in the first part of the proof that there is a measure $\gamma^E \in \mathcal{M}(\overline{E})$ such that $\gamma_n \to \gamma^E$ weakly, and then that $\mathcal{E}_{k,\nu,p}(\gamma^E - \gamma_n) \to 0$, as

$n \to \infty$. This then implies that a subsequence $\{V_{k,\nu,p}^{\gamma_{n_j}}\}_{j=1}^\infty$ converges to $V_{k,\nu,p}^{\gamma^E}$ on $\mathbf{R}^N \setminus A$, where $C_{k,\nu,p}(A) = 0$. It follows as before that (3.2.9) holds. Now let $x \in \operatorname{supp} \gamma^E \setminus A$. Then

$$V_{k,\nu,p}^{\gamma^E}(x) = \lim_{j\to\infty} V_{k,\nu,p}^{\gamma_{n_j}}(x) \leq \lim_{j\to\infty} \mathcal{E}_{k,\nu,p}(\gamma_{n_j}) = \mathcal{E}_{k,\nu,p}(\gamma^E),$$

where the inequality is a consequence of the fact that $\operatorname{supp} \gamma^E = \bigcap_{n=1}^\infty \operatorname{supp} \gamma_n$. Finally,

$$\mathcal{E}_{k,\nu,p}(\gamma^E) = \lim_{n\to\infty} \mathcal{E}_{k,\nu,p}(\gamma_n) = \lim_{n\to\infty} C_{k,\nu,p}(G_n)^{-p'/p} = C_{k,\nu,p}(E)^{-p'/p}. \quad \square$$

3.3. Bessel and Riesz capacities

It is well-known that Bessel and Riesz capacities occur naturally in the study of the deeper properties of the Sobolev space $W_w^{m,p}(\mathbf{R}^N)$ for $1 < p < \infty$, for instance, as tools for measuring exceptional sets for functions belonging to this space. For a full account of this and the history of the subject, see Adams–Hedberg [7], Maz'ya [76], or Ziemer [108]. Weighted analogues of such capacities have been investigated by, among others, D. R. Adams [5], [6], H. Aikawa [10], and T. Kilpeläinen [67]. One reason for the usefulness of Bessel capacities when studying Sobolev spaces is the fact that these capacities are equivalent to different variational capacities that are associated with the norm in $W_w^{m,p}(\mathbf{R}^N)$. Below we show that such an equivalence also holds in the weighted context. We will also prove that the weighted Bessel capacity is comparable to an inhomogeneous Riesz capacity, which has the advantage of being easier to work with than the Bessel capacity. Adams calculated in [5] the weighted Riesz capacity of a ball. Here, we give a more elementary proof of Adams' result, and also generalize it to include inhomogeneous Riesz capacities.

3.3.1. Basic properties

We begin by defining the weighted Bessel capacity of a set.

Definition 3.3.1. Let $0 < \alpha < N$ and $1 < p < \infty$. Let w be a weight. If $E \subset \mathbf{R}^N$, we define the *weighted Bessel capacity* of E, $B_{\alpha,p}^w(E)$, by

$$B_{\alpha,p}^w(E) = \inf\{\|f\|_{L_w^p(\mathbf{R}^N)}^p \; ; \; f \in L_w^p(\mathbf{R}^N)^+ \text{ and } \mathcal{G}_\alpha f(x) \geq 1 \text{ for every } x \in E\}.$$

The principal motivation for our interest in Bessel capacities comes from the theorem below, which shows that the space of Bessel potentials of order m of functions in $L_w^p(\mathbf{R}^N)$ coincides with the Sobolev space $W_w^{m,p}(\mathbf{R}^N)$, if w is an A_p weight, and that the spaces have equivalent norms. Without weights, this is a classical result by A. P. Calderón [18]. The weighted version was proved by N. Miller [82, p. 104]. Later, E. Nieminen [91, p. 38] rediscovered Miller's result and gave an elegant proof of the theorem, new even in the non-weighted case.

Theorem 3.3.2. *Let m be a positive integer, and let $1 < p < \infty$. Let w be an A_p weight. Suppose that $f \in L_w^p(\mathbf{R}^N)$. Then the function $u = \mathcal{G}_m f$ belongs to $W_w^{m,p}(\mathbf{R}^N)$. Furthermore, there are positive constants C_1 and C_2, that only depend on m, N, p, and the A_p constant of w, such that*

$$C_1 \|f\|_{L_w^p(\mathbf{R}^N)} \leq \|u\|_{W_w^{m,p}(\mathbf{R}^N)} \leq C_2 \|f\|_{L_w^p(\mathbf{R}^N)}.$$

Conversely, every function u, belonging to $W_w^{m,p}(\mathbf{R}^N)$, may be written $u = \mathcal{G}_m f$ for some function $f \in L_w^p(\mathbf{R}^N)$.

With the aid of this theorem, we will show that the Bessel capacity $B_{m,p}^w$ on compact subsets of $\mathbf{R}^N$ is equivalent to the Sobolev capacity $c_{m,p}^w$, a capacity naturally connected to the space $W_w^{m,p}(\mathbf{R}^N)$. We postpone the proof to Section 3.5, where we have collected some results concerning variational capacities.

Definition 3.3.3. Let m be an integer, $1 \leq m < N$, and $1 < p < \infty$, and let w be a weight. If $K \subset \mathbf{R}^N$ is compact, we define the *weighted, m-th order Sobolev capacity* $c_{m,p}^w(K)$ of K by

$$c_{m,p}^w(K) = \inf\{ \|\varphi\|_{W_w^{m,p}(\mathbf{R}^N)}^p ; \varphi \in C_0^\infty(\mathbf{R}^N) \text{ and } \varphi \geq 1 \text{ on } K\}.$$

Theorem 3.3.4. *Let m be an integer, $1 \leq m < N$, and let $1 < p < \infty$. Let w be an A_p weight. Then there are two positive constants, C_1 and C_2, that only depend on m, N, p, and the A_p constant of w, such that*

$$C_1 c_{m,p}^w(K) \leq B_{m,p}^w(K) \leq C_2 c_{m,p}^w(K) \tag{3.3.1}$$

for every compact set $K \subset \mathbf{R}^N$.

We now turn our attention to the Riesz capacities $R_{\alpha,p;\rho}^w$. Due to the relative simplicity of the Riesz kernel in comparison with the Bessel kernel, it is often preferable to work with various equivalent Riesz capacities rather than directly with the Bessel capacity.

Definition 3.3.5. Let $0 < \alpha < N$, $1 < p < \infty$, and $0 < \rho \leq \infty$. Let w be a weight. If $E \subset \mathbf{R}^N$, we define the *weighted Riesz capacity* $R_{\alpha,p;\rho}^w(E)$ of E by

$$R_{\alpha,p;\rho}^w(E) = \inf\{\|f\|_{L_w^p(\mathbf{R}^N)}^p ; f \geq 0 \text{ and } \mathcal{I}_{\alpha,\rho} f(x) \geq 1 \text{ for every } x \in E\}.$$

Remark 3.3.6. We will write $R_{\alpha,p;\rho}(E)$ instead of $R_{\alpha,p;\rho}^1(E)$ and $R_{\alpha,p}^w(E)$ instead of $R_{\alpha,p;\infty}^w(E)$.

After some preliminary considerations, our first task will be to show that the inhomogeneous capacity $R_{\alpha,p;1}^w$ is equivalent to $B_{\alpha,p}^w$. Notice that this fact together with Theorem 3.3.4 implies that, for integer α, $R_{\alpha,p;1}^w$ also is equivalent to the Sobolev capacity $c_{\alpha,p}^w$.

The fact that the the weighted Riesz capacity $R_{\alpha,p;\rho}^w$ is comprised by Meyers' theory in Section 3.2 will be of great importance. To see this, we note as in Adams [5, p. 79] that if we let

$$k(x,y) = I_{\alpha,\rho}(x-y)w(y)^{-1}$$

for $x, y \in \mathbf{R}^N$, and define $d\nu(y) = w(y)\,dy$, then

$$C_{k,\nu,p}(E) = R_{\alpha,p;\rho}^w(E).$$

Thus, the general theory of capacities applies to $R_{\alpha,p;\rho}^w$. The nonlinear potential of a measure μ on $\mathbf{R}^N$ is

$$V_\rho^\mu(x) = \mathcal{I}_{\alpha,\rho}((\mathcal{I}_{\alpha,\rho}\mu)^{p'-1}w')(x),$$

where, as always, $w' = w^{-1/(p-1)}$. Note that, by (3.2.2), the nonlinear energy of μ may be written

$$\int_{\mathbf{R}^N} V_\rho^\mu \, d\mu = \int_{\mathbf{R}^N} (\mathcal{I}_{\alpha,\rho}\mu)^{p'} w' \, dx. \qquad (3.3.2)$$

In exactly the same way, one shows that the capacity $B_{\alpha,p}^w$ is a special case of the general theory. The nonlinear energy of a measure is in this case

$$\int_{\mathbf{R}^N} (\mathcal{G}_{\alpha,\rho}\mu)^{p'} w' \, dx.$$

Theorem 3.3.7. *Let $0 < \alpha < N$ and $1 < p < \infty$. Let w be an A_p weight. Then there are positive constants C_1 and C_2 such that if $E \subset \mathbf{R}^N$, then*

$$C_1 R_{\alpha,p;1}^w(E) \le B_{\alpha,p}^w(E) \le C_2 R_{\alpha,p;1}^w(E). \qquad (3.3.3)$$

The constants C_1 and C_2 depend only on α, N, p, and the A_p constant of w.

Lemma 3.3.8. *Let $0 < \alpha < N$, $1 < p < \infty$, and $0 < \rho \le \infty$. Let w be an A_p weight. Then there exists a positive constant C such that if $E \subset \mathbf{R}^N$, then*

$$R_{\alpha,p;\rho}^w(E) \le C R_{\alpha,p;2\rho}^w(E). \qquad (3.3.4)$$

The constant C depends only on α, N, p, and the A_p constant of w.

Proof. Let $\mu \in \mathcal{M}^+(\mathbf{R}^N)$. According to Corollary 3.1.4,

$$\int_{\mathbf{R}^N} (\mathcal{I}_{\alpha,2\rho}\mu)^{p'} w' \, dx \le C \int_{\mathbf{R}^N} (M_{\alpha,\rho}\mu)^{p'} w' \, dx. \qquad (3.3.5)$$

Now, it is well-known and easy to see that

$$M_{\alpha,\rho}\mu(x) \le \mathcal{I}_{\alpha,\rho}\mu(x) \qquad (3.3.6)$$

for every $x \in \mathbf{R}^N$. By inserting this in the right-hand side of the inequality (3.3.5), we obtain

$$\int_{\mathbf{R}^N} (\mathcal{I}_{\alpha,2\rho}\mu)^{p'} w' \, dx \le C \int_{\mathbf{R}^N} (\mathcal{I}_{\alpha,\rho}\mu)^{p'} w' \, dx.$$

This inequality together with the dual definitions (Proposition 3.2.3) imply that

$$R^w_{\alpha,p;\rho}(K) \le C R^w_{\alpha,p;2\rho}(K). \tag{3.3.7}$$

for any compact set K. If we now use the facts that both capacities are continuous from the left and that every open set is the union of an increasing sequence of compact sets, (3.3.7) extends to open sets. The general case then follows from Proposition 3.2.4. □

Proof of Theorem 3.3.7. The second inequality in (3.3.3) follows from the asymptotic formula (3.1.9) and the previous lemma. For the proof of the first inequality, we use Corollary 3.1.5, which (in the case $\rho = 1$) states that

$$\int_{\mathbf{R}^N} (\mathcal{G}_\alpha\mu)^{p'} w' \, dx \le C \int_{\mathbf{R}^N} (M_{\alpha,1}\mu)^{p'} w' \, dx$$

for every positive measure μ. If we combine this inequality with (3.3.6), we see that

$$\int_{\mathbf{R}^N} (\mathcal{G}_\alpha\mu)^{p'} w' \, dx \le C \int_{\mathbf{R}^N} (\mathcal{I}_{\alpha,1}\mu)^{p'} w' \, dx,$$

and hence

$$R^w_{\alpha,p;1}(K) \le C B^w_{\alpha,p}(K)$$

for any compact set K. This inequality then extends to general sets as in the proof of Lemma 3.3.8. □

Notice that Lemma 3.3.8 implies that if $0 < \rho_1 < \rho_2 < \infty$, then the capacities $R^w_{\alpha,p;\rho_1}$ and $R^w_{\alpha,p;\rho_2}$ are comparable since evidently $R^w_{\alpha,p;\rho_2}(E) \le R^w_{\alpha,p;\rho_1}(E)$ for every set $E \subset \mathbf{R}^N$.

In this connection, we would like to mention a similar result by H. Aikawa (see [10, pp. 343–344]), which we will have occasion to return to in Section 3.4.

Theorem 3.3.9. *Let $0 < \alpha < N$ and $1 < p < \infty$. Let w be an A_p weight.*

(a) *$R^w_{\alpha,p}$ is nontrivial, i.e., not identically 0, if and only if $w \in A_{p,\alpha}$, where $A_{p,\alpha}$ is the set of A_p weights such that*

$$\int_{\mathbf{R}^N} \frac{w'(x)}{(1+|x|)^{(N-\alpha)p'}} \, dx < \infty. \tag{3.3.8}$$

(b) *If $w \in A_{p,\alpha}$, then $R^w_{\alpha,p}$ is locally equivalent to $B^w_{\alpha,p}$ in the sense that if $E \subset B_R(a)$, then there is a positive constant C, depending only on a, α, N, p, R, and the A_p constant of w, such that*

$$R^w_{\alpha,p}(E) \le B^w_{\alpha,p}(E) \le C R^w_{\alpha,p}(E).$$

3.3.2. Adams' formula for the capacity of a ball

We now come to the formula for the Riesz capacity of a ball, which, in the homogeneous case $\rho = \infty$, was established by D. R. Adams [5, pp. 82–83]. With $p = 2$ and $w \in A_2$, similar results had previously been obtained by E. B. Fabes, D. S. Jerison, and C. E. Kenig [35]. Adams' result was later extended to more general kernels by S. K. Vodop'yanov in [104].

Much more general results have recently been obtained by N. J. Kalton and I. E. Verbitsky [63]. They consider the capacity $C_{k,\nu,p}$ (see Section 3.2), where ν is an arbitrary measure and the kernel k is assumed to be positive, symmetric, and satisfy a so called quasi-metric condition, and have established two-sided estimates for the capacity of a ball. Their result gives Adams' formula with absolutely no restrictions on the weight. Unfortunately it does not cover the inhomogeneous Riesz capacity.[†]

Our proof of Adams' formula uses only the definition of Riesz capacity and is consequently longer and more complicated than Adams' proof.[‡]

Theorem 3.3.10. *Let* $1 < p < \infty$, $0 < \alpha < N$, *and* $0 < \rho \leq \infty$. *Let* w *be a weight such that* $w' = w^{-1/(p-1)}$ *is doubling. In the case* $\rho = \infty$, *suppose that the condition* (3.3.8) *above is satisfied. Then, if* $a \in \mathbf{R}^N$ *and* $0 < r \leq \rho$ $(0 < r < \infty, \ if \ \rho = \infty)$,

$$R^w_{\alpha,p;\rho}(B_r(a)) \leq C \left(\int_r^{2\rho} \frac{w'(B_t(a))}{t^{(N-\alpha)p'}} \frac{dt}{t} \right)^{1-p}, \tag{3.3.9}$$

where the constant C *only depends on* N, α, p, *and the doubling constant of* w. *If* $w \in A_p$, *the inequality* (3.3.9) *may be reversed with a constant that then, in addition, also depends on the* A_p *constant of* w.

Remark 3.3.11.

1. Besides guaranteeing that $R^w_{\alpha,p}$ is nontrivial, the condition (3.3.8) implies that the integral in the right-hand side of (3.3.9) is convergent.

2. Techniques, resembling the ones in the second part of the proof, were used by Aikawa in his proof of the local equivalence between Bessel and Riesz capacities, i.e., part (b) of Theorem 3.3.9.

We begin by proving a lemma.

Lemma 3.3.12. *Let* $0 < \alpha < N$, $1 < p < \infty$, *and* $0 < \rho < \infty$. *Let* w *be an* A_p *weight. Then there are positive constants* C_1 *and* C_2, *that only depend on* N, α, p, *and the* A_p *constant of* w, *such that*

$$C_1 \rho^{-\alpha p} w(B_\rho(a)) \leq R^w_{\alpha,p;\rho}(B_\rho(a)) \leq C_2 \rho^{-\alpha p} w(B_\rho(a))$$

[†]One expects however that Kalton's and Verbitsky's approach can be modified to include arbitrary radially decreasing kernels, perhaps with some restrictions on the measure.

[‡]Adams uses Wolff's inequality, Theorem 3.6.6.

for every ball $B_\rho(a)$.

Proof. We first prove the upper bound. Set $f = A\chi_{B_\rho(a)}w'$, where the value of the constant A will be specified later. We then have, for $x \in B_\rho(a)$,

$$\mathcal{I}_{\alpha,\rho}f(x) = A\int_{|x-y|<\rho,|a-y|<\rho} \frac{w'(y)}{|x-y|^{N-\alpha}}\,dy \geq A\rho^{-(N-\alpha)}w'(E),$$

where E is the set $\{y\;;\;|x-y| < \rho, |a-y| < \rho\}$. Also, by the strong doubling property of w' (recall that, by Remark 1.2.4.4, $w' \in A_{p'}$),

$$w'(E) \geq Cw'(B_\rho(a)).$$

Thus, by choosing $A = \rho^{N-\alpha}C^{-1}w'(B_\rho(a))^{-1}$, it follows that $\mathcal{I}_{\alpha,\rho}f \geq 1$ on the ball $B_\rho(a)$. Moreover,

$$\int_{\mathbf{R}^N} f^p w\,dx = A^p w'(B_\rho(a)) = \rho^{(N-\alpha)p}C^{-p}w'(B_\rho(a))^{-p}w'(B_\rho(a))$$

$$= C\rho^{N-\alpha p}\left(\fint_{B_\rho(a)} w'\,dx\right)^{1-p}.$$

If we now use Hölder's inequality (cf. Remark 1.2.4.2), we obtain

$$\int_{\mathbf{R}^N} f^p w\,dx \leq C\rho^{-\alpha p}w(B_\rho(a)).$$

This gives the upper bound for $R^w_{\alpha,p;\rho}(B_\rho(a))$.

For the lower bound, we let $\varepsilon > 0$ be arbitrary, and then let $f \geq 0$ satisfy $\mathcal{I}_{\alpha,\rho}f \geq 1$ on $B_\rho(a)$ and

$$\int_{\mathbf{R}^N} f^p w\,dx < R^w_{\alpha,p;\rho}(B_\rho(a)) + \varepsilon.$$

The second assumption concerning f implies that

$$w(B_\rho(a)) \leq \int_{B_\rho(a)} (\mathcal{I}_{\alpha,\rho}f)^p w\,dx.$$

It is easily verified that

$$\mathcal{I}_{\alpha,\rho}f(x) \leq C\rho^\alpha Mf(x)$$

(cf. the identity (2.4.6)), so according to Muckenhoupt's maximal inequality (Theorem 1.2.3),

$$\int_{B_\rho(a)} (\mathcal{I}_{\alpha,\rho}f)^p w\,dx \leq C\rho^{\alpha p}\int_{\mathbf{R}^N} f^p w\,dx.$$

It follows that

$$\rho^{-\alpha p} w(B_\rho(a)) \le C \int_{\mathbf{R}^N} f^p w \, dx < R^w_{\alpha,p;\rho}(B_\rho(a)) + \varepsilon.$$

This proves the lemma since ε was arbitrary. $\square$

Proof of Theorem 3.3.10. In the proof of the inequality (3.3.9), we will assume that $\rho < \infty$, the case $\rho = \infty$ being analogous, but somewhat simpler. Let $C_1 = 2^{N-\alpha} D^4 / ((N-\alpha) p')$, where D is the doubling constant of w', and $f = C_1 A^{-1} g$, where

$$A = \int_r^{2\rho} \frac{w'(B_t(a))}{t^{(N-\alpha)p'}} \frac{dt}{t},$$

and

$$g(x) = \begin{cases} r^{-(N-\alpha)(p'-1)} w'(x), & \text{if } |x-a| < r, \\ |x-a|^{-(N-\alpha)(p'-1)} w'(x), & \text{if } r \le |x-a| < 2\rho, \text{ and} \\ 0, & \text{otherwise.} \end{cases}$$

Then, according to the identity (2.4.6),

$$\int_{\mathbf{R}^N} |g|^p w \, dx = \frac{w'(B_r(a))}{r^{(N-\alpha)p'}} + (N-\alpha)p' \int_r^{2\rho} \frac{w'(B_t(a))}{t^{(N-\alpha)p'}} \frac{dt}{t}$$
$$+ \frac{w'(B_{2\rho}(a))}{(2\rho)^{(N-\alpha)p'}} - \frac{w'(B_r(a))}{r^{(N-\alpha)p'}} \le CA.$$

Here we have used the fact that

$$\int_r^{2\rho} \frac{w'(B_t(a))}{t^{(N-\alpha)p'}} \frac{dt}{t} \ge \int_\rho^{2\rho} \frac{w'(B_t(a))}{t^{(N-\alpha)p'}} \frac{dt}{t} \ge C \frac{w'(B_\rho(a))}{(2\rho)^{(N-\alpha)p'}}$$
$$\ge \frac{C}{D} \frac{w'(B_{2\rho}(a))}{(2\rho)^{(N-\alpha)p'}}. \tag{3.3.10}$$

Suppose that $x \in B_r(a)$. Then

$$\mathcal{I}_{\alpha,\rho} g(x) = r^{-(N-\alpha)(p'-1)} \int_{|y-a|<r, |x-y|<\rho} \frac{w'(y)}{|x-y|^{N-\alpha}} \, dy$$
$$+ \int_{r \le |y-a|<2\rho, |x-y|<\rho} \frac{w'(y)}{|x-y|^{N-\alpha}|y-a|^{(N-\alpha)(p'-1)}} \, dy$$
$$\ge r^{-(N-\alpha)(p'-1)} \frac{w'(B_r(a) \cap B_\rho(x))}{(2r)^{N-\alpha}}$$
$$+ 2^{-(N-\alpha)} \int_{r \le |y-a|<2\rho} \frac{\chi_{B_\rho(x)}(y) w'(y)}{|y-a|^{(N-\alpha)p'}} \, dy$$

$$= 2^{-(N-\alpha)} \frac{w'(B_r(a) \cap B_\rho(x))}{r^{(N-\alpha)p'}}$$

$$+ (N-\alpha)p'2^{-(N-\alpha)} \int_r^{2\rho} \frac{w'(B_t(a) \cap B_\rho(x))}{t^{(N-\alpha)p'}} \frac{dt}{t}$$

$$+ 2^{-(N-\alpha)} \left(\frac{w'(B_{2\rho}(a) \cap B_\rho(x))}{(2\rho)^{(N-\alpha)p'}} - \frac{w'(B_r(a) \cap B_\rho(x))}{r^{(N-\alpha)p'}} \right)$$

$$\geq 2^{-(N-\alpha)}(N-\alpha)p' \int_r^{2\rho} \frac{w'(B_t(a) \cap B_\rho(x))}{t^{(N-\alpha)p'}} \frac{dt}{t}.$$

We now claim that

$$w'(B_t(a) \cap B_\rho(x)) \geq D^{-4} w'(B_t(a)) \tag{3.3.11}$$

for $r \leq t \leq 2\rho$. This is obvious when $t \leq \frac{1}{2}\rho$ (and consequently $r \leq \frac{1}{2}\rho$), since then $B_t(a) \subset B_\rho(x)$. Suppose $\frac{1}{2}\rho \leq t \leq 2\rho$. If we set $z = (3a+x)/4$, then $B_t(a) \cap B_\rho(x) \supset B_{\rho/4}(z)$ and $B_{4\rho}(z) \supset B_t(a)$. The doubling property of w' now implies

$$w'(B_t(a) \cap B_\rho(x)) \geq w'(B_{\rho/4}(z)) \geq D^{-4} w'(B_{4\rho}(z)) \geq D^{-4} w'(B_t(a)),$$

which proves (3.3.11). Using this fact and the estimates above, we find

$$\mathcal{I}_{\alpha,\rho} g(x) \geq 2^{-(N-\alpha)} D^{-4} (N-\alpha)p' \int_r^{2\rho} \frac{w'(B_t(a))}{t^{(N-\alpha)p'}} \frac{dt}{t} = \frac{A}{C_1}.$$

It follows that

$$\mathcal{I}_{\alpha,\rho} f(x) = \frac{C_1}{A} \mathcal{I}_{\alpha,\rho} g(x) \geq 1$$

on $B_r(a)$. Furthermore, by the estimates of the norm of g,

$$R_{\alpha,p;\rho}^w(B_r(a)) \leq \int_{\mathbf{R}^N} f^p w \, dx = \frac{C_1^p}{A^p} \int_{\mathbf{R}^N} g^p w \, dx \leq \frac{C_1^p}{A^p} CA$$

$$= CC_1^p \left(\int_r^{2\rho} \frac{w'(B_t(a))}{t^{(N-\alpha)p'}} \frac{dt}{t} \right)^{1-p},$$

which is (3.3.9).

We now prove the reverse inequality, and assume as before that $\rho < \infty$. Let $\varepsilon > 0$ be arbitrary, and let $f \geq 0$ satisfy $\mathcal{I}_{\alpha,\rho} f \geq 1$ on $B_r(a)$ and

$$\int_{\mathbf{R}^N} f^p w \, dx < R_{\alpha,p;\rho}^w(B_r(a)) + \varepsilon.$$

Let $E = \{x \in B_r(a) \, ; \, \mathcal{I}_{\alpha,r} f(x) < \frac{1}{2}\}$. If $E \neq \emptyset$, then for some $x \in B_r(a)$,

$$\frac{1}{2} \leq \int_{r \leq |x-y| \leq \rho} \frac{f(y)}{|x-y|^{N-\alpha}} \, dy$$

$$\leq \left(\int_{\mathbf{R}^N} f^p w \, dx \right)^{1/p} \left(\int_{r \leq |x-y| \leq \rho} \frac{w'(y)}{|x-y|^{(N-\alpha)p'}} \, dy \right)^{1/p'}. \tag{3.3.12}$$

Another application of the identity (2.4.6) now shows that

$$\int_{r \le |x-y| \le \rho} \frac{w'(y)}{|x-y|^{N-\alpha}} \, dy = (N-\alpha)p' \int_r^\rho \frac{w'(B_t(x))}{t^{(N-\alpha)p'}} \frac{dt}{t}$$
$$+ \frac{w'(B_\rho(x))}{\rho^{(N-\alpha)p'}} - \frac{w'(B_r(x))}{r^{(N-\alpha)p'}}.$$

If $t \ge r$, then $B_t(x) \subset B_{2t}(a)$, so $w'(B_t(x)) \le w'(B_{2t}(a))$. If we use this fact and the inequality (3.3.10) once more, we obtain

$$\int_{r \le |x-y| \le \rho} \frac{w'(y)}{|x-y|^{N-\alpha}} \, dy \le C \left(\int_r^\rho \frac{w'(B_{2t}(a))}{t^{(N-\alpha)p'}} \frac{dt}{t} + \int_\rho^{2\rho} \frac{w'(B_t(a))}{t^{(N-\alpha)p'}} \frac{dt}{t} \right)$$
$$= C \int_r^{2\rho} \frac{w'(B_t(a))}{t^{(N-\alpha)p'}} \frac{dt}{t}.$$

By inserting this in (3.3.12), it follows that

$$\frac{1}{2} \le C \left(R_{\alpha,p;\rho}^w(B_r(a)) + \varepsilon \right)^{1/p} \left(\int_r^{2\rho} \frac{w'(B_t(a))}{t^{(N-\alpha)p'}} \frac{dt}{t} \right)^{(p-1)/p},$$

and finally,

$$R_{\alpha,p;\rho}^w(B_r(a)) \ge C \left(\int_r^{2\rho} \frac{w'(B_t(a))}{t^{(N-\alpha)p'}} \frac{dt}{t} \right)^{1-p}.$$

If $E = \emptyset$, then $I_{\alpha,r} * 2f \ge 1$ on $B_r(a)$, so

$$R_{\alpha,p;r}^w(B_r(a)) \le \int_{\mathbf{R}^N} 2f^p w \, dx \le 2^p (R_{\alpha,p;\rho}^w(B_r(a)) + \varepsilon),$$

which implies that

$$R_{\alpha,p;\rho}^w(B_r(a)) \ge 2^{-p} R_{\alpha,p;r}^w(B_r(a)).$$

We now use the fact that $R_{\alpha,p;r}^w(B_r(a))$, by Lemma 3.3.12, is comparable to $r^{-\alpha p} w(B_r(a))$. Also, by Remark 1.2.4.2,

$$\frac{w(B_r(a))}{r^{\alpha p}} \ge \left(\frac{w'(B_r(a))}{r^{(N-\alpha)p'}} \right)^{1-p},$$

and since

$$\int_r^{2\rho} \frac{w'(B_t(a))}{t^{(N-\alpha)p'}} \frac{dt}{t} \ge \int_r^{2r} \frac{w'(B_t(a))}{t^{(N-\alpha)p'}} \frac{dt}{t} \ge C \frac{w'(B_r(a))}{r^{(N-\alpha)p'}},$$

the lower bound for $R_{\alpha,p;\rho}^w(B_r(a))$ follows. $\square$

3.4. Hausdorff capacities

This section is devoted to weighted Hausdorff capacities.[†] These capacities will be used in later sections for measuring exceptional sets for functions belonging to the space $W_w^{m,1}(\Omega)$. Also, in Section 3.7 below, we suggest a definition of thinness based on this capacity. In the present section, we intend to establish counterparts for Hausdorff capacities to the properties already proved in Section 3.3 for Bessel and Riesz capacities. To begin with, we calculate the weighted Hausdorff capacity of a ball. We then characterize the weights for which the homogeneous capacity is nontrivial. Following this, the continuity of the Hausdorff capacity with respect to the ordinary limit processes is established and used to prove a capacitability theorem. Our final result is a weighted version of a classical theorem by O. Frostman [42].[‡]

3.4.1. Basic properties

Let us begin by recalling some notation from Section 2.3. Let w be a weight, and let $0 \leq \alpha < N$ and $0 < \rho \leq \infty$. For a ball $B_r(a)$, we set

$$h_w^{N-\alpha}(B_r(a)) = \frac{w(B_r(a))}{r^\alpha}.$$

Let E be a subset of $\mathbf{R}^N$. We then define the *weighted, $(N-\alpha)$-dimensional Hausdorff capacity* of E, $\mathcal{H}_{w,\rho}^{N-\alpha}(E)$, by

$$\mathcal{H}_{w,\rho}^{N-\alpha}(E) = \inf \sum_j h_w^{N-\alpha}(B_j),$$

where the infimum is taken over all countable coverings of E with open balls B_j such that each ball has radius $\leq \rho$ (in the case $\rho < \infty$). If $w = 1$, we shall write $\mathcal{H}_\rho^{N-\alpha}$ instead of $\mathcal{H}_{1,\rho}^{N-\alpha}$.

We next define a dyadic version of $\mathcal{H}_{w,\rho}^{N-\alpha}$. By a *dyadic cube* we shall mean a cube

$$Q = 2^{-j}(k + [0,1)^N) = \{2^{-j}(k+y) \; ; \; y \in [0,1)^N\},$$

where $j \in \mathbf{Z}$ and $k \in \mathbf{Z}^N$. If Q is dyadic with side-length l, we set

$$h_w^{N-\alpha}(Q) = \frac{w(Q)}{l^\alpha}.$$

We then define the *dyadic, weighted, $(N-\alpha)$-dimensional Hausdorff capacity* of E, $\widetilde{\mathcal{H}}_{w,\rho}^{N-\alpha}(E)$, by

$$\widetilde{\mathcal{H}}_{w,\rho}^{N-\alpha}(E) = \inf \sum_j h_w^{N-\alpha}(Q_j),$$

[†]Another name for Hausdorff capacity is *Hausdorff content*.

[‡]Hausdorff capacities have occurred earlier in the work of for instance D. R. Adams [4], L. Carleson [19], A. Carlsson [20], [21], A. Carlsson and V. G. Maz'ya [22], C. Fernström [40], Yu. V. Netrusov [90].

where the infimum is taken over all countable coverings of E with dyadic cubes Q_j such that each Q_j has side-length $\leq \rho$ (when $\rho < \infty$) and E is *properly covered* by $\{Q_j\}$, i.e., $E \subset (\bigcup_j Q_j)^\circ$. We will call such coverings *proper*.

Our first object is to show that these set functions are equivalent when w is doubling. For the proof of this fact, we need a lemma, that will enable us to compare Hausdorff capacities $\mathcal{H}_{w,\rho}^{N-\alpha}$ with different (finite) bounds for the radii in the admissible coverings. This lemma should be compared with Proposition 3.3.8 in Section 3.3.1. In Proposition 3.4.15 below, we address the question under what conditions the capacities $\mathcal{H}_{w,\rho}^{N-\alpha}$, $0 < \rho < \infty$, and $\mathcal{H}_{w,\infty}^{N-\alpha}$ are locally equivalent.

Lemma 3.4.1. *Let $0 \leq \alpha < N$ and $0 < \rho < \sigma < \infty$. Let w be a doubling weight. Then there is a positive constant C, which only depends on α, N, and the doubling constant of w, such that, for every set $E \subset \mathbf{R}^N$,*

$$\mathcal{H}_{w,\sigma}^{N-\alpha}(E) \leq \mathcal{H}_{w,\rho}^{N-\alpha}(E) \leq C \left(\frac{\sigma}{\rho}\right)^\alpha \mathcal{H}_{w,\sigma}^{N-\alpha}(E). \tag{3.4.1}$$

Proof. The inequality to the left in (3.4.1) is obvious. For the proof of the second inequality, we let $\varepsilon > 0$ be arbitrary, and let $\{B_j\}$ be a covering of E such that each ball B_j has radius $r_j \leq \sigma$ and

$$\sum_j h_w^{N-\alpha}(B_j) < \mathcal{H}_{w,\sigma}^{N-\alpha}(E) + \varepsilon.$$

Every B_j can be covered by $m = m(N, \rho/\sigma)$ balls $B_k^{(j)}$, $k = 1, \ldots, m$, such that each ball $B_k^{(j)}$ has radius $r_k^{(j)} = \rho r_j / \sigma$ and the multiplicity of the covering only depends on N. Then $B_k^{(j)} \subset 3B_j$ for every k and

$$\sum_{k=1}^m h_w^{N-\alpha}(B_k^{(j)}) = \left(\frac{\sigma}{\rho}\right)^\alpha r_j^{-\alpha} \sum_{k=1}^m w(B_k^{(j)}) \leq C \left(\frac{\sigma}{\rho}\right)^\alpha r_j^{-\alpha} w(3B_j)$$

$$\leq C \left(\frac{\sigma}{\rho}\right)^\alpha h_w^{N-\alpha}(B_j),$$

where the last inequality follows from the doubling property of w. Note that $\{B_k^{(j)}\}_{j,k}$ is an admissible covering of E for $\mathcal{H}_{w,\rho}^{N-\alpha}$. Moreover, by the inequality above,

$$\mathcal{H}_{w,\rho}^{N-\alpha}(E) \leq \sum_{j,k} h_w^{N-\alpha}(B_k^{(j)}) = \sum_j \sum_{k=1}^m h_w^{N-\alpha}(B_k^{(j)})$$

$$\leq C \left(\frac{\sigma}{\rho}\right)^\alpha \sum_j h_w^{N-\alpha}(B_j) < C \left(\frac{\sigma}{\rho}\right)^\alpha (\mathcal{H}_{w,\sigma}^{N-\alpha}(E) + \varepsilon).$$

This proves the second inequality in (3.4.1). $\square$

Proposition 3.4.2. *Let* $0 \leq \alpha < N$ *and* $0 < \rho \leq \infty$. *If* w *is doubling, then there are positive constants* C_1 *and* C_2, *which only depend on* α, N, *and the doubling constant of* w, *such that, for every set* $E \subset \mathbf{R}^N$,

$$C_1 \mathcal{H}_{w,\rho}^{N-\alpha}(E) \leq \widetilde{\mathcal{H}}_{w,\rho}^{N-\alpha}(E) \leq C_2 \mathcal{H}_{w,\rho}^{N-\alpha}(E). \tag{3.4.2}$$

Proof. Suppose that Q is a dyadic cube with center a and side-length l. Then $B_{l/2}(a) \subset Q \subset B_{l\sqrt{N}}(a)$. The doubling property of w now shows that

$$\frac{w(B_{l\sqrt{N}}(a))}{(l\sqrt{N})^\alpha} \leq C \frac{w(B_{l/2}(a))}{l^\alpha} \leq C \frac{w(Q)}{l^\alpha}. \tag{3.4.3}$$

In the case $\rho = \infty$, this implies that $\mathcal{H}_{w,\infty}^{N-\alpha}(E) \leq \widetilde{\mathcal{H}}_{w,\infty}^{N-\alpha}(E)$ for every set E, and if $0 < \rho < \infty$, it follows from (3.4.3) and the preceding lemma that

$$\mathcal{H}_{w,\rho}^{N-\alpha}(E) \leq C \mathcal{H}_{w,\rho\sqrt{N}}^{N-\alpha}(E) \leq C \widetilde{\mathcal{H}}_{w,\rho}^{N-\alpha}(E)$$

for every $E \subset \mathbf{R}^N$.

For the proof of the reverse inequality, we note that every ball $B_r(a)$, with $2^{-j} \leq r < 2^{-j+1}$, can be covered by maximally 2^{2N} dyadic cubes Q_k with side-lengths 2^{-j}. Since $Q_k \subset B_{3r}(a)$ for every k, we obtain

$$\sum_{k=1}^{2^{2N}} \frac{w(Q_k)}{2^{-\alpha j}} \leq 2^{2N} 2^{\alpha j} w(B_{3r}(a)) \leq C \frac{w(B_r(a))}{r^\alpha},$$

from which the second inequality in (3.4.3) follows. $\square$

We shall frequently use the following properties, already mentioned in Section 2.3, of the set functions $\mathcal{H}_{w,\rho}^{N-\alpha}$ and $\widetilde{\mathcal{H}}_{w,\rho}^{N-\alpha}$; cf. Remark 2.3.4.

Proposition 3.4.3. *Let* $0 \leq \alpha < N$ *and* $0 < \rho \leq \infty$. *We then have*

(a) $\mathcal{H}_{w,\rho}^{N-\alpha}(\emptyset) = 0$;

(b) *if* $E_1 \subset E_2$, *then* $\mathcal{H}_{w,\rho}^{N-\alpha}(E_1) \leq \mathcal{H}_{w,\rho}^{N-\alpha}(E_2)$;

(c) $\mathcal{H}_{w,\rho}^{N-\alpha}\left(\bigcup_{n=1}^\infty E_n\right) \leq \sum_{n=1}^\infty \mathcal{H}_{w,\rho}^{N-\alpha}(E_n)$.

These properties also hold for $\widetilde{H}_{w,\rho}^{N-\alpha}$.

3.4.2. The capacity of a ball

We will next give estimates for the $\mathcal{H}_{w,\rho}^{N-\alpha}$ capacity of a ball, and begin by proving the following useful lemma.

Lemma 3.4.4. *Let $0 \leq \alpha < N$ and $0 < \rho \leq \infty$, and suppose that $w \in A_1$. Then there are positive constants C_1 and C_2, depending only on α, N, and the A_1 constant of w, such that if $0 < r \leq \rho$ and $B_s(y) \subset B_r(x)$, then*

$$C_1 \left(\frac{s}{r}\right)^{N-\alpha} \inf_{r \leq t \leq \rho} \frac{w(B_t(x))}{t^\alpha} \leq \inf_{s \leq t \leq \rho} \frac{w(B_t(y))}{t^\alpha} \leq C_2 \inf_{r \leq t \leq \rho} \frac{w(B_t(x))}{t^\alpha}. \quad (3.4.4)$$

It follows that if $B_s(y) \subset B_r(x)$ (with no restrictions on r), then

$$C_1 \left(\frac{s}{r}\right)^{N-\alpha} \inf_{t \geq r} \frac{w(B_t(x))}{t^\alpha} \leq \inf_{t \geq s} \frac{w(B_t(y))}{t^\alpha} \leq C_2 \inf_{t \geq r} \frac{w(B_t(x))}{t^\alpha}.$$

Proof. Let us first prove the second of the two inequalities in (3.4.4). Let t satisfy $r \leq t \leq \rho$, and let $B_\sigma(y)$ be the smallest ball with center y such that $B_\sigma(y) \supset B_t(x)$, i.e., take $\sigma = t + |x - y|$. The strong doubling property of w then implies that

$$\frac{w(B_t(x))}{t^\alpha} \geq C \left(\frac{t}{t + |x - y|}\right)^{N-\alpha} \frac{w(B_\sigma(y))}{\sigma^\alpha} \geq C 2^{-(N-\alpha)} \frac{w(B_\sigma(y))}{\sigma^\alpha}. \quad (3.4.5)$$

Note that $r \leq \sigma \leq \rho + r$. If $r \leq \sigma \leq \rho$, the desired inequality follows immediately from (3.4.5), and if $\rho \leq \sigma \leq \rho + r \leq 2\rho$, we have

$$\frac{w(B_\sigma(y))}{\sigma^\alpha} \geq C \left(\frac{\sigma}{2\rho}\right)^{N-\alpha} \frac{w(B_{2\rho}(y))}{(2\rho)^\alpha} \geq C \frac{w(B_\rho(y))}{\rho^\alpha}, \quad (3.4.6)$$

and we are done.

We now turn to the proof of the first inequality in (3.4.4). Let $s \leq t \leq \rho$ and set $\sigma = \max\{r, t + |x - y|\}$. Then $r \leq \sigma \leq \rho + r$ and $B_\sigma(x) \supset B_t(y)$. We then find

$$\frac{w(B_t(y))}{t^\alpha} \geq C \left(\frac{t}{\sigma}\right)^{N-\alpha} \frac{w(B_\sigma(x))}{\sigma^\alpha}.$$

To finish the proof, we have to show that

$$\left(\frac{t}{\sigma}\right)^{N-\alpha} \frac{w(B_\sigma(x))}{\sigma^\alpha} \geq C \left(\frac{s}{r}\right)^{N-\alpha} \inf_{r \leq t \leq \rho} \frac{w(B_t(x))}{t^\alpha}. \quad (3.4.7)$$

This is obvious when $\sigma = r$ because $t \geq s$. Suppose that $\sigma = t + |x - y|$. Then

$$\frac{t}{\sigma} = \frac{t}{t + |x - y|} \geq \frac{s}{s + |x - y|} \geq \frac{s}{2r}.$$

The inequality (3.4.7) now follows in the same way as in the first part of the proof. $\square$

Corollary 3.4.5. *Let $0 \leq \alpha < N$, and suppose that $w \in A_1$. Then there are positive constants C_1 and C_2, depending only on α, N, and the A_1 constant of w, such that, for every x and y in $\mathbf{R}^N$,*

$$C_1 \liminf_{r \to \infty} \frac{w(B_r(x))}{r^\alpha} \leq \liminf_{r \to \infty} \frac{w(B_r(y))}{r^\alpha} \leq C_2 \liminf_{r \to \infty} \frac{w(B_r(x))}{r^\alpha}.$$

Proof. This follows from the second set of inequalities in Proposition 3.4.4 if we first let $s > 0$ be arbitrary, then take $r = s + |x - y|$, and finally let $s \to \infty$. □

Theorem 3.4.6. *Let $0 \leq \alpha < N$ and $0 < \rho \leq \infty$, and suppose that $w \in A_1$. If $\rho < \infty$, then, for every ball $B_r(a)$ with radius $r \leq \rho$,*

$$C_1 \inf_{r \leq t \leq \rho} \frac{w(B_t(a))}{t^\alpha} \leq \mathcal{H}_{w,\rho}^{N-\alpha}(B_r(a)) \leq \inf_{r \leq t \leq \rho} \frac{w(B_t(a))}{t^\alpha}.$$

If $\rho = \infty$, then, for every ball $B_r(a)$,

$$C_2 \inf_{t \geq r} \frac{w(B_t(a))}{t^\alpha} \leq \mathcal{H}_{w,\infty}^{N-\alpha}(B_r(a)) \leq \inf_{t \geq r} \frac{w(B_t(a))}{t^\alpha}.$$

The constants C_1 and C_2, appearing in these inequalities, depend only on α, N, and the A_1 constant of w.

Proof. We will give the proof in the case $\rho < \infty$. The upper bound is immediate since $B_r(a)$ is covered by $B_t(a)$, $r \leq t \leq \rho$. To obtain the lower bound, we let $\varepsilon > 0$ be arbitrary and choose a covering $\{B_j\}$ of $B_r(a)$ with balls B_j such that if r_j is the radius of B_j, then $r_j \leq \rho$ for every j, and

$$\sum_j h_w^{N-\alpha}(B_j) < \mathcal{H}_{w,\rho}^{N-\alpha}(B_r(a)) + \varepsilon.$$

First, suppose that $r_j \leq r$ for every j. Then $B_j \subset B_{3r}(a)$, so by the strong doubling property of w,

$$h_w^{N-\alpha}(B_j) = \frac{w(B_j)}{r_j^\alpha} \geq C \left(\frac{r_j}{r}\right)^{N-\alpha} \frac{w(B_r(a))}{r^\alpha},$$

and we get

$$\sum_j h_w^{N-\alpha}(B_j) \geq C \frac{w(B_r(a))}{r^\alpha} \frac{1}{r^{N-\alpha}} \sum_j r_j^{N-\alpha}.$$

If we now use the fact that $\mathcal{H}_\rho^{N-\alpha}(B_r(a)) = Cr^{N-\alpha}$ (see Fernström [40, p. 25]), it follows that

$$\sum_j r_j^{N-\alpha} \geq \mathcal{H}_\rho^{N-\alpha}(B_r(a)) = Cr^{N-\alpha},$$

whence

$$\mathcal{H}_{w,\rho}^{N-\alpha}(B_r(a)) + \varepsilon \geq C\frac{w(B_r(a))}{r^\alpha},$$

which gives the lower bound for $\mathcal{H}_{w,\rho}^{N-\alpha}(B_r(a))$ in the case $r_j \leq r$ for every j.

Now suppose $r_{j_0} > r$ for some j_0. Since we obviously may assume that B_{j_0} intersects $B_r(a)$, we have $B_r(a) \subset 3B_{j_0}$. Lemma 3.4.4 and the doubling property of w now imply that

$$\inf_{r \leq t \leq 3\rho} \frac{w(B_t(a))}{t^\alpha} \leq C\frac{w(B_{3r_{j_0}}(x_{j_0}))}{(3r_{j_0})^\alpha} \leq C\frac{w(B_{r_{j_0}}(x_{j_0}))}{r_{j_0}^\alpha} \leq C\sum_j h_w^{N-\alpha}(B_j),$$

and hence

$$\mathcal{H}_{w,\rho}^{N-\alpha}(B_r(a)) + \varepsilon > C\inf_{r \leq t \leq 3\rho} \frac{w(B_t(a))}{t^\alpha}.$$

But if $\rho \leq t \leq 3\rho$, then as in (3.4.6),

$$\frac{w(B_t(a))}{t^\alpha} \geq C\frac{w(B_\rho(a))}{\rho^\alpha},$$

which implies that

$$\inf_{r \leq t \leq 3\rho} \frac{w(B_t(a))}{t^\alpha} \geq C\inf_{r \leq t \leq \rho} \frac{w(B_t(a))}{t^\alpha}.$$

This completes the proof. □

Remark 3.4.7.

1. The upper bound for $\mathcal{H}_{w,\rho}^{N-\alpha}(B_r(a))$ holds for arbitrary weights w.

2. Note that $\mathcal{H}_{w,\infty}^{N-\alpha}(B_r(a))$ may be 0. We will return to this phenomenon shortly.

Here follows some observations concerning Theorem 3.4.6. Let us begin by comparing the estimates for $\mathcal{H}_{w,\rho}^{N-\alpha}(B_r(a))$ with the results in Theorem 3.3.10. We know, for instance, that if $0 < r \leq 1$ and $w \in A_p$, then $R_{\alpha,p;1}^w(B_r(a))$ is comparable to

$$\left(\int_r^2 \frac{w'(B_t(a))}{t^{(N-\alpha)p'}} \frac{dt}{t}\right)^{1-p}.$$

By the A_p condition and Remark 1.2.4.2, an equivalent expression is

$$\left(\int_r^2 \left(\frac{t^{\alpha p}}{w(B_t(a))}\right)^{1/(p-1)} \frac{dt}{t}\right)^{1-p}. \tag{3.4.8}$$

As $p \to 1$, the quantity (3.4.8) tends to $\inf_{r \leq t \leq 2} t^{-\alpha}w(B_t(a))$, which in turn can be compared with

$$\inf_{r \leq t \leq 1} \frac{w(B_t(a))}{t^\alpha},$$

that is, the quantity that appears in the estimates for $\mathcal{H}^{N-\alpha}_{w,1}(B_r(a))$. It is reasonable to expect the capacities $R^w_{\alpha,p;1}$ and $\mathcal{H}^{N-\alpha}_{w,1}$ to have similar properties since, if α is an integer, then by Theorem 3.3.4, $R^w_{\alpha,p;1}$ is equivalent to the variational capacity $c^w_{\alpha,p}$ and, as will be shown in Theorem 3.5.5, $\mathcal{H}^{N-\alpha}_{w,1}$ is equivalent to the variational capacity $C^w_{\alpha,1}$.

Let $0 < r \leq \rho < \infty$. It is obvious, and also follows from the proposition, that, for $x \in \mathbf{R}^N$,

$$\mathcal{H}^{N-\alpha}_{w,\rho}(B_r(x)) \leq \frac{w(B_r(x))}{r^\alpha}. \tag{3.4.9}$$

To study the reverse inequality, we prove a proposition.

Proposition 3.4.8. *Let $0 \leq \alpha < N$ and $0 < \rho < \infty$, and let w be a weight. Then*

$$\lim_{r \to 0} \frac{r^\alpha}{w(B_r(x))} \inf_{r \leq t \leq \rho} \frac{w(B_t(x))}{t^\alpha} = 1. \tag{3.4.10}$$

for a.e. $x \in \mathbf{R}^N$.

Proof. Evidently

$$\inf_{r \leq t \leq \rho} \frac{w(B_t(x))}{t^\alpha} \leq \frac{w(B_r(x))}{r^\alpha},$$

so that

$$\limsup_{r \to 0} \frac{r^\alpha}{w(B_r(x))} \inf_{r \leq t \leq \rho} \frac{w(B_t(x))}{t^\alpha} \leq 1.$$

Let x be such that $0 < w(x) < \infty$ and

$$\fint_{B_t(x)} w \, dy \to w(x), \quad \text{as } t \to 0.$$

For an arbitrary $\varepsilon > 0$, we may then choose t_0, $0 < t_0 < \rho$, so that

$$\frac{1}{1+\varepsilon} w(x) \leq \fint_{B_t(x)} w \, dy \leq (1+\varepsilon)w(x)$$

for $0 < t < t_0$. Let $0 < r \leq t_0$. We have

$$\frac{r^\alpha}{w(B_r(x))} \inf_{r \leq t \leq \rho} \frac{w(B_t(x))}{t^\alpha} = \left(\frac{w(B_r(x))}{r^\alpha} \sup_{r \leq t \leq \rho} \frac{t^\alpha}{w(B_t(x))} \right)^{-1},$$

and, furthermore,

$$\frac{w(B_r(x))}{r^\alpha} \sup_{r \leq t \leq \rho} \frac{t^\alpha}{w(B_t(x))}$$

$$\leq Cr^{N-\alpha}(1+\varepsilon)w(x) \left(\sup_{r \leq t \leq t_0} \frac{t^\alpha}{w(B_t(x))} + \sup_{t_0 \leq t \leq \rho} \frac{t^\alpha}{w(B_t(x))} \right)$$

$$\leq Cr^{N-\alpha}(1+\varepsilon)w(x) \left(\frac{1+\varepsilon}{Cw(x)} r^{-(N-\alpha)} + \sup_{t_0 \leq t \leq \rho} \frac{t^\alpha}{w(B_t(x))} \right)$$

$$\leq (1+\varepsilon)^2 + (1+\varepsilon)\varepsilon = 1 + 3\varepsilon + 2\varepsilon^2,$$

where the last inequality holds if r is small enough, $0 < r < r_0$, say. Thus

$$\frac{r^\alpha}{w(B_r(x))} \inf_{r \leq t \leq \rho} \frac{w(B_t(x))}{t^\alpha} \geq \frac{1}{1 + 3\varepsilon + 2\varepsilon^2}$$

for $0 < r < r_0$, from which it follows that

$$\liminf_{r \to 0} \frac{r^\alpha}{w(B_r(x))} \inf_{r \leq t \leq \rho} \frac{w(B_t(x))}{t^\alpha} \geq 1. \quad \square$$

The inequality (3.4.9) may thus essentially be reversed a.e., as $r \to 0$. Note that this implies that $\mathcal{H}_{w,\rho}^{N-\alpha}(B_r(x))$ locally behaves like $Cr^{N-\alpha}w(x)$ a.e. These results also hold in the case $\rho = \infty$, if we assume that

$$\inf_{t \geq r} \frac{w(B_t(x))}{t^\alpha} > 0$$

for a.e. $x \in \mathbf{R}^N$ and every $r > 0$ (cf. Lemma 3.4.13).

Let $\overline{B_r(x)}$ be a closed ball with radius $r \leq \rho < \infty$. Evidently

$$\mathcal{H}_{w,\rho}^{N-\alpha}(\overline{B_r(x)}) \leq \inf_{r \leq t \leq \rho} \frac{w(B_t(x))}{t^\alpha},$$

and by Proposition 3.4.3(b) and the preceding proposition,

$$\mathcal{H}_{w,\rho}^{N-\alpha}(\overline{B_r(x)}) \geq \mathcal{H}_{w,\rho}^{N-\alpha}(B_r(x)) \geq C \inf_{r \leq t \leq \rho} \frac{w(B_t(x))}{t^\alpha}.$$

Using Proposition 3.4.23 below, we see that $\mathcal{H}_{w,\rho}^{N-\alpha}(\overline{B_r(x)}) \to \mathcal{H}_{w,\rho}^{N-\alpha}(\{x\})$, as $r \to 0$. Hence

$$C_1 \inf_{0 < t \leq \rho} \frac{w(B_t(x))}{t^\alpha} \leq \mathcal{H}_{w,\rho}^{N-\alpha}(\{x\}) \leq C_2 \inf_{0 < t \leq \rho} \frac{w(B_t(x))}{t^\alpha}.$$

Simple examples show that a point x may have positive capacity, e.g., if the weight has a singularity at x.

3.4.3. Non-triviality of $\mathcal{H}_{w,\infty}^{N-\alpha}$

Before proving any further properties of our capacities, we have to exclude some "pathological" cases, namely, weights such that $\mathcal{H}_{w,\infty}^{N-\alpha}$ is identically 0 (and, consequently, the same holds for $\widetilde{\mathcal{H}}_{w,\infty}^{N-\alpha}$) and weights with the property that $\widetilde{\mathcal{H}}_{w,\infty}^{N-\alpha}$ is not continuous from the left (cf. Proposition 3.4.22). It is more or less obvious that $\mathcal{H}_{w,\infty}^{N-\alpha}$ and $\widetilde{\mathcal{H}}_{w,\infty}^{N-\alpha}$ can behave strangely, if the weight w is "too small at infinity," thus making the measure function $h_w^{N-\alpha}(B_r(x))$ decreasing, or at least bounded, as a function of r. We will first characterize those A_1 weights w for which $\mathcal{H}_{w,\infty}^{N-\alpha}$ is identically 0.

Definition 3.4.9. For $0 \le \alpha < N$, let $A_{1,\alpha}$ be the set of A_1 weights w for which there exists a positive constant C such that

$$w(x) \ge \frac{C}{(1 + |x|)^{N-\alpha}}$$

for a.e. $x \in \mathbf{R}^N$.

Remark 3.4.10. According to Remark 1.2.4.8, for every weight $w \in A_1$, there exists a constant $C > 0$ such that $w(x) \ge C(1 + |x|)^{-N}$ a.e. Thus, $A_{1,0} = A_1$.

Theorem 3.4.11. Let $0 \le \alpha < N$, and suppose that $w \in A_1$. Then

(a) $\mathcal{H}^{N-\alpha}_{w,\infty}(\mathbf{R}^N) = \infty$, if $w \in A_{1,\alpha}$, and

(b) $\mathcal{H}^{N-\alpha}_{w,\infty}(\mathbf{R}^N) = 0$, if $w \notin A_{1,\alpha}$.

Remark 3.4.12. Let $1 < p < N/\alpha$. Recall that, by Theorem 3.3.9(a), a result due to H. Aikawa,

$$R^w_{\alpha,p}(\mathbf{R}^N) = \begin{cases} \infty, & \text{if } w \in A_{p,\alpha}, \\ 0, & \text{if } w \notin A_{p,\alpha}, \end{cases}$$

where $w \in A_{p,\alpha}$, if $w \in A_p$ and

$$\left(\int_{\mathbf{R}^N} \frac{w'(x)}{(1 + |x|)^{(N-\alpha)p'}}\, dx \right)^{p-1} < \infty. \tag{3.4.11}$$

The condition defining $A_{1,\alpha}$ is, at least formally, the limit of (3.4.11), as $p \to 1$.

For the proof of Theorem 3.4.11, we need a lemma and a proposition.

Lemma 3.4.13. Let $0 \le \alpha < N$, and suppose that $w \in A_1$. Then $w \in A_{1,\alpha}$ if and only if there exists a constant $C > 0$ such that

$$\frac{w(B_r(a))}{r^\alpha} \ge C \tag{3.4.12}$$

for every $a \in \mathbf{R}^N$ and every $r \ge 1 + |a|$.

Proof. First, suppose that $w \in A_{1,\alpha}$, i.e., there exists a constant $C > 0$ such that

$$w(x) \ge \frac{C}{(1 + |x|)^{N-\alpha}}$$

for a.e. $x \in \mathbf{R}^N$. Let $a \in \mathbf{R}^N$ and $r \ge 1 + |a|$. Then

$$w(B_r(a)) \ge Cw(B_{2r}(a)) \ge Cw(B_r(0)) \ge C \int_{|y|<r} \frac{dy}{(1 + |y|)^{N-\alpha}}$$

$$\ge C \frac{r^N}{(1 + r)^{N-\alpha}} \ge Cr^\alpha,$$

thus proving (3.4.12).

To prove the converse, let $\{B_i\}_{i=1}^{\infty}$ be an enumeration of the rational balls in $\mathbf{R}^N$. By the A_1 condition, we have, for $i = 1, 2, \dots$,

$$w(x) \geq C \fint_{B_i} w \, dy$$

for a.e. $x \in B_i$. Let $E_i \subset B_i$ be the exceptional set and define $E = \bigcup_{i=1}^{\infty} E_i$. Then $|E| = 0$. Let $x \in \mathbf{R}^N \setminus E$. Let $z \in \mathbf{Q}^N$ satisfy $|x - z| \leq |x|$, and let $q \in \mathbf{Q}$ be such that $|x| \leq q \leq 2|x|$. Then $x \in B_{1+q}(z)$, and we obtain

$$w(x) \geq C \fint_{B_{1+q}(z)} w \, dy \geq \frac{C}{(1+q)^{N-\alpha}} \geq \frac{C}{(1+|x|)^{N-\alpha}},$$

that is, $w \in A_{1,\alpha}$. $\square$

Proposition 3.4.14. Let $0 \leq \alpha < N$, and suppose that $w \in A_1$. Then

(a) $w \in A_{1,\alpha}$ if and only if there exists a constant $C > 0$ such that

$$\liminf_{r \to \infty} \frac{w(B_r(a))}{r^\alpha} \geq C \tag{3.4.13}$$

for every $a \in \mathbf{R}^N$;

(b) $w \notin A_{1,\alpha}$ if and only if

$$\liminf_{r \to \infty} \frac{w(B_r(a))}{r^\alpha} = 0 \tag{3.4.14}$$

for every $a \in \mathbf{R}^N$.

Proof. (a) In view of Lemma 3.4.13, the necessity of (3.4.13) for $w \in A_{1,\alpha}$ is obvious; we will prove the sufficiency by showing that condition (3.4.12) in the lemma is fulfilled. Choose $t_0 \geq 2$ so large that

$$\frac{w(B_t(0))}{t^\alpha} \geq \frac{C}{2}$$

for $t \geq t_0$. Let $a \in \mathbf{R}^N$ and $r \geq 1 + |a|$ be arbitrary. Set $t = r + |a|$. Suppose $t \geq t_0$. The strong doubling property of w then implies that

$$\frac{w(B_r(a))}{r^\alpha} \geq C_1 \left(\frac{r}{t} \right)^{N-\alpha} \frac{w(B_t(0))}{t^\alpha} \geq C_1 2^{-(N-\alpha)} \frac{C}{2} \geq C_1 t_0^{-(N-\alpha)} \frac{C}{2}.$$

If $t < t_0$, we find in the same way that

$$\frac{w(B_r(a))}{r^\alpha} \geq C_1 \left(\frac{r}{t_0} \right)^{N-\alpha} \frac{w(B_{t_0}(0))}{t_0^\alpha} \geq C_1 t_0^{-(N-\alpha)} \frac{C}{2}.$$

(b) If $w \notin A_{1,\alpha}$, then $\liminf_{r\to\infty} r^{-\alpha} w(B_r(a))$ cannot be positive for any $a \in \mathbf{R}^N$, since this, according to Corollary 3.4.5, would imply the existence of a constant $C > 0$ such that $\liminf_{r\to\infty} r^{-\alpha} w(B_r(a)) \geq C$ for every $a \in \mathbf{R}^N$, and thus, by part (a), that $w \in A_{1,\alpha}$. The converse follows directly from part (a). $\square$

We now come to the proof of Theorem 3.4.11.

Proof of Theorem 3.4.11. (a) We will show that if $\{B_j\}_{j=1}^\infty$ is a covering of $\mathbf{R}^N$ with balls $B_j = B_{r_j}(x_j)$, then

$$\sum_{j=1}^\infty h_w^{N-\alpha}(B_j) = \infty. \qquad (3.4.15)$$

According to Lemma 3.4.13, there is a constant $C_1 > 0$ such that

$$\frac{w(B_r(a))}{r^\alpha} \geq C_1 \qquad (3.4.16)$$

for every $a \in \mathbf{R}^N$ and every $r \geq 1 + |a|$. It follows that if $r_j \geq 1 + |x_j|$ for an infinite number of j, then the series in (3.4.15) will be automatically divergent. We may therefore assume that $r_j < 1 + |x_j|$ for all but a finite number of j, and, after removing a finite number of balls, that this holds for every j and

$$\bigcup_{j=1}^\infty B_j = \mathbf{R}^N \setminus E, \qquad (3.4.17)$$

where E is bounded. Then, by the strong doubling property of w and (3.4.16),

$$\frac{w(B_j)}{r_j^\alpha} \geq C \left(\frac{r_j}{1+|x_j|} \right)^{N-\alpha} \frac{w(B_{1+|x_j|}(x_j))}{(1+|x_j|)^\alpha} \geq CC_1 \left(\frac{|B_j|}{(1+|x_j|)^N} \right)^{(N-\alpha)/N}.$$

If we now use the inequality (2.4.16), it follows that

$$\sum_{j=1}^\infty \frac{w(B_j)}{r_j^\alpha} \geq C \sum_{j=1}^\infty \left(\frac{|B_j|}{(1+|x_j|)^N} \right)^{(N-\alpha)/N}$$

$$\geq C \left(\sum_{j=1}^\infty \frac{|B_j|}{(1+|x_j|)^N} \right)^{(N-\alpha)/N}. \qquad (3.4.18)$$

If $|x_j| < 1$ for some j, then $B_j \subset B_2(0)$ since $r_j < 1 + |x_j|$, so we can assume that $|x_j| \geq 1$ for every j and that (3.4.17) still holds. We may also assume that every r_j is less than $\frac{1}{2}|x_j|$ because the series in the right-hand side of (3.4.18) diverges if $r_j \geq \frac{1}{2}|x_j|$ for an infinite number of j. Note that this implies that

$$\frac{|B_j|}{(1+|x_j|)^N} \geq 2^{-N} \int_{B_j} \frac{dx}{(1+|x|)^N},$$

whence

$$\sum_{j=1}^{\infty} \frac{|B_j|}{(1+|x_j|)^N} \geq 2^{-N} \sum_{j=1}^{\infty} \int_{B_j} \frac{dx}{(1+|x|)^N} \geq 2^{-N} \int_{\mathbf{R}^N \setminus E} \frac{dx}{(1+|x|)^N}.$$

This proves (3.4.15) since the last integral is divergent.

(b) Because $w \notin A_{1,\alpha}$, we have by Proposition 3.4.14,

$$\liminf_{r \to \infty} \frac{w(B_r(a))}{r^\alpha} = 0$$

for every $a \in \mathbf{R}^N$. Let $\varepsilon > 0$ be arbitrary and choose for every $k \in \mathbf{Z}^N$ a number $r_k > \sqrt{N}$ such that

$$\frac{w(B_{r_k}(k))}{r_k^\alpha} < \varepsilon |k|^{-2N}.$$

Then $\{B_{r_k}(k)\}_{k \in \mathbf{Z}^N}$ covers $\mathbf{R}^N$ and

$$\mathcal{H}^{N-\alpha}_{w,\infty}(\mathbf{R}^N)S \leq \sum_{k \in \mathbf{Z}^N} \frac{w(B_{r_k}(k))}{r_k^\alpha} < \sum_{k \in \mathbf{Z}^N} \varepsilon |k|^{-2N} = C\varepsilon.$$

The claim now follows because ε was arbitrary. $\square$

3.4.4. Local equivalence between $\mathcal{H}^{N-\alpha}_{w,\rho}$ and $\mathcal{H}^{N-\alpha}_{w,\infty}$

Proposition 3.4.15 gives sufficient conditions for $\mathcal{H}^{N-\alpha}_{w,\rho}$, $0 < \rho < \infty$, to be locally equivalent to $\mathcal{H}^{N-\alpha}_{w,\infty}$ (cf. Theorem 3.3.9).

Proposition 3.4.15. *Let $0 \leq \alpha < N$ and $0 < \rho < \infty$, and let $w \in A_{1,\alpha}$. Suppose that E is a subset of a ball $B_R(a)$. Then*

$$\mathcal{H}^{N-\alpha}_{w,\infty}(E) \leq \mathcal{H}^{N-\alpha}_{w,\rho}(E) \leq C\mathcal{H}^{N-\alpha}_{w,\infty}(E), \tag{3.4.19}$$

where the constant C is given by

$$C = C_1 \frac{\mathcal{H}^{N-\alpha}_{w,\rho}(B_R(a))}{\mathcal{H}^{N-\alpha}_{w,\infty}(B_R(a))} \max\{1, (R/\rho)^\alpha\}$$

and C_1 only depends on α, N, and the A_1 constant of w.

Proof. We only have to prove the second inequality in (3.4.19). Let $\varepsilon > 0$ be arbitrary, and choose a covering of E with balls B_j such that

$$\sum_j h^{N-\alpha}_w(B_j) < \mathcal{H}^{N-\alpha}_{w,\infty}(E) + \varepsilon$$

and such that every B_j intersects E. Let r_j be the radius of B_j. First, suppose that $r_j < R$ for every j. Then

$$\mathcal{H}_{w,R}^{N-\alpha}(E) \leq \sum_j h_w^{N-\alpha}(B_j) < \mathcal{H}_{w,\infty}^{N-\alpha}(E) + \varepsilon,$$

and thus, $\mathcal{H}_{w,R}^{N-\alpha}(E) \leq \mathcal{H}_{w,\infty}^{N-\alpha}(E)$. This proves the inequality to the right in (3.4.19), since, if $\rho < R$, Lemma 3.4.1 implies that

$$\mathcal{H}_{w,R}^{N-\alpha}(E) \geq C \left(\frac{\rho}{R}\right)^\alpha \mathcal{H}_{w,\rho}^{N-\alpha}(E) \geq C \frac{\mathcal{H}_{w,\infty}^{N-\alpha}(B_R(a))}{\mathcal{H}_{w,\rho}^{N-\alpha}(B_R(a))} \left(\frac{\rho}{R}\right)^\alpha \mathcal{H}_{w,\rho}^{N-\alpha}(E),$$

and if $\rho \geq R$, then

$$\mathcal{H}_{w,R}^{N-\alpha}(E) \geq \mathcal{H}_{w,\rho}^{N-\alpha}(E) \geq \frac{\mathcal{H}_{w,\infty}^{N-\alpha}(B_R(a))}{\mathcal{H}_{w,\rho}^{N-\alpha}(B_R(a))} \mathcal{H}_{w,\rho}^{N-\alpha}(E).$$

Now suppose that $r_j \geq R$ for some j. Note that $B_R(a) \subset 3B_j$, so according to Lemma 3.4.4 and Theorem 3.4.6,

$$\mathcal{H}_{w,\infty}^{N-\alpha}(E) + \varepsilon > h_w^{N-\alpha}(B_j) \geq C h_w^{N-\alpha}(3B_j) \geq C \inf_{t \geq R} \frac{w(B_t(a))}{t^\alpha}$$

$$\geq C \mathcal{H}_{w,\infty}^{N-\alpha}(B_R(a)) = C \frac{\mathcal{H}_{w,\infty}^{N-\alpha}(B_R(a))}{\mathcal{H}_{w,\rho}^{N-\alpha}(B_R(a))} \mathcal{H}_{w,\rho}^{N-\alpha}(B_R(a))$$

$$\geq C \frac{\mathcal{H}_{w,\infty}^{N-\alpha}(B_R(a))}{\mathcal{H}_{w,\rho}^{N-\alpha}(B_R(a))} \min \{1, (\rho/R)^\alpha\} \mathcal{H}_{w,\rho}^{N-\alpha}(E),$$

and (3.4.19) follows. $\square$

Using this lemma, we can prove the following proposition.

Proposition 3.4.16. *Let* $0 \leq \alpha < N$ *and* $0 < \rho \leq \infty$, *and let* w *be an* A_1 *weight. In the case* $\rho = \infty$, *suppose that* $w \in A_{1,\alpha}$. *If* $E \subset \mathbf{R}^N$ *with* $\mathcal{H}_{w,\rho}^{N-\alpha}(E) = 0$, *then* $\mathcal{H}_\rho^{N-\alpha}(E) = 0$.

Remark 3.4.17. It follows from well-known properties of Hausdorff measures (see for instance Evans–Gariepy [34, pp. 64–65]) and from the proposition that $\mathcal{H}^N(E) = |E| = 0$, if $\mathcal{H}_{w,\rho}^{N-\alpha}(E) = 0$; this is, in fact, what we shall use later.

Proof. By Proposition 3.4.3(c), we may assume that E is a subset of a ball $B_R(a)$, and by Lemma 3.4.1 and Lemma 3.4.15, that $\rho = R$. Let $\varepsilon > 0$ be arbitrary, and cover E with balls B_j such that each B_j has radius $r_j \leq R$, $B_j \cap E \neq \emptyset$ for every j, and $\sum_j h_w^{N-\alpha}(B_j) < \varepsilon$. Then every B_j is included in $B_{3R}(a)$, and we have

$$h_w^{N-\alpha}(B_j) \geq C \left(\frac{r_j}{R}\right)^{N-\alpha} \frac{w(B_R(a))}{R^\alpha} \geq C \frac{\mathcal{H}_{w,R}^{N-\alpha}(B_R(a))}{\mathcal{H}_R^{N-\alpha}(B_R(a))} r_j^{N-\alpha}.$$

Summing over j, we obtain

$$\mathcal{H}_R^{N-\alpha}(E) \leq C \frac{\mathcal{H}_R^{N-\alpha}(B_R(a))}{\mathcal{H}_{w,R}^{N-\alpha}(B_R(a))} \varepsilon,$$

and since ε was arbitrary, this implies that $\mathcal{H}_R^{N-\alpha}(E) = 0$. $\quad\square$

Remark 3.4.18. Essentially the same argument as in this proof shows that, if $E \subset B_R(a)$, then

$$\mathcal{H}_R^{N-\alpha}(E) \leq C \frac{\mathcal{H}_R^{N-\alpha}(B_R(a))}{\mathcal{H}_{w,R}^{N-\alpha}(B_R(a))} \mathcal{H}_{w,R}^{N-\alpha}(E).$$

By combining this inequality with Lemma 3.4.1 and Lemma 3.4.15, it follows that $\mathcal{H}_\rho^{N-\alpha}(E) \leq C \mathcal{H}_{w,\rho}^{N-\alpha}(E)$ for $0 < \rho \leq \infty$ (assuming that $w \in A_{1,\alpha}$ in the case $\rho = \infty$).

3.4.5. Continuity properties

We will next investigate to what extent $\widetilde{\mathcal{H}}_{w,\rho}^{N-\alpha}$ is continuous from the left, and begin by considering an example.

Example 3.4.19. The weight $w(x) = |x|^{-(\alpha-N)}, 0 < \alpha \leq N$, belongs to $A_{1,\alpha}$, and hence $\widetilde{\mathcal{H}}_{w,\infty}^{N-\alpha}(\mathbf{R}^N) = \infty$. But

$$C_1 r^\alpha \leq w(B_r(a)) \leq C_2 r^\alpha$$

for $r \geq |a|$, which implies that $\widetilde{\mathcal{H}}_{w,\infty}^{N-\alpha}(B_r(a))$ is essentially constant for large r. It follows that $\widetilde{\mathcal{H}}_{w,\infty}^{N-\alpha}(\mathbf{R}^N) \neq \lim_{r\to\infty} \widetilde{\mathcal{H}}_{w,\infty}^{N-\alpha}(B_r(a))$ for any $a \in \mathbf{R}^N$. In other words, $\widetilde{H}_{w,\infty}^{N-\alpha}$ is not continuous from the left.

The situation in this example should be compared with the limiting case $\alpha = N$ for $w = 1$, where $\mathbf{R}^N$ has infinite capacity but every ball has capacity $|B_1(0)|$. The reason for the problem which occurs in the example is the fact that the measure function $h_w^{N-\alpha}(B_r(x))$ is bounded from above as a function of r. By Remark 3.4.21.1 and Proposition 3.4.22 below, the class $A_{1,\alpha}^+$, which we define next, consists precisely of those A_1 weights w for which $\widetilde{\mathcal{H}}_{w,\infty}^{N-\alpha}$ is continuous from the left.

Definition 3.4.20. For $0 \leq \alpha < N$, let $A_{1,\alpha}^+$ be the set of A_1 weights w for which

$$\lim_{r\to\infty} \frac{w(B_r(a))}{r^\alpha} = \infty \tag{3.4.20}$$

for every $a \in \mathbf{R}^N$.

Remark 3.4.21.

1. It follows from Theorem 3.4.6, that if (3.4.20) does not hold and $w \in A_{1,\alpha}$, then $\widetilde{H}_{w,\infty}^{N-\alpha}$ is not continuous from the left.

2. By Proposition 3.4.14, $A_{1,\alpha}^+ \subset A_{1,\alpha}$.

3. $A_{1,0}^+ = A_1$, since $\lim_{r\to\infty} w(B_r(a)) = \int_{\mathbf{R}^N} w\, dx = \infty$ for any $a \in \mathbf{R}^N$ (see Remark 1.2.11).

The proof of the proposition below is adapted from Carleson [19, pp. 9–10].

Proposition 3.4.22. *Let $0 \leq \alpha < N$ and $0 < \rho \leq \infty$. Let w be a doubling weight. In the case $\rho = \infty$, assume that*

$$\lim_{r\to\infty} \frac{w(B_r(a))}{r^\alpha} = \infty \qquad (3.4.21)$$

for every $a \in \mathbf{R}^N$. If $E_1 \subset E_2 \subset \dots$ is an increasing sequence of subsets of $\mathbf{R}^N$, then

$$\widetilde{\mathcal{H}}_{w,\rho}^{N-\alpha}\left(\bigcup_{n=1}^{\infty} E_n\right) = \lim_{n\to\infty} \widetilde{\mathcal{H}}_{w,\rho}^{N-\alpha}(E_n).$$

Proof. Note that the limit $l = \lim_{n\to\infty} \widetilde{\mathcal{H}}_{w,\rho}^{N-\alpha}(E_n)$ exists. If $l = \infty$, the proposition is trivial, so suppose $l < \infty$. Let $E = \bigcup_{n=1}^{\infty} E_n$. It then suffices to prove that $\widetilde{\mathcal{H}}_{w,\rho}^{N-\alpha}(E) \leq l$. To this end, we let $\varepsilon > 0$ be arbitrary and choose for each n a proper, dyadic covering $\{Q_j^{(n)}\}_j$ of E_n, satisfying

$$\sum_j h_w^{N-\alpha}(Q_j^{(n)}) < \widetilde{\mathcal{H}}_{w,\rho}^{N-\alpha}(E_n) + 2^{-n}\varepsilon. \qquad (3.4.22)$$

We now claim that, for every $x \in E$, there is a maximal cube $Q_x \in \{Q_j^{(n)}\}_{j,n}$ such that $x \in Q_x$. This is obvious when $\rho < \infty$, so we will consider the case $\rho = \infty$. Suppose that, for some $x \in E$, there exists an infinite chain $x \in Q_1 \subset Q_2 \subset \dots$ of different dyadic cubes $Q_m \in \{Q_j^{(n)}\}_{j,n}$. If a is the center of Q_1 and l_m is the side-length of Q_m, then $Q_m \subset B_{l_m\sqrt{N}}(a)$, and hence

$$h_w^{N-\alpha}(Q_m) = \frac{w(Q_m)}{l_m^\alpha} \geq C\frac{w(B_{l_m\sqrt{N}}(a))}{(l_m\sqrt{N})^\alpha}.$$

By assumption, the right-hand side of the last inequality tends to ∞, as $m \to \infty$. But this is impossible since (3.4.22) implies that $h_w^{N-\alpha}(Q_m) < l + \varepsilon$ for every cube $Q_j^{(n)}$ and especially for the cubes Q_m. This proves the claim. The maximal cubes Q_x form a countable, proper covering $\{Q_i\}$ of E.

Now let $m \geq 1$ be fixed and consider the cubes Q_i, $i \in M_1$, such that $Q_i \in \{Q_j^{(1)}\}$. These cubes cover a subset F_1 of E_m. The set F_1 is also covered by cubes $Q_j^{(m)}$, $j \in N_1$. The cubes Q_i, $i \in M_1$, and $Q_j^{(m)}$, $j \in N_1$, cover $F_1 \cap E_1$, so according to (3.4.22),

$$\sum_{i \in M_1} h_w^{N-\alpha}(Q_i) \leq \sum_{j \in N_1} h_w^{N-\alpha}(Q_j^{(m)}) + 2^{-1}\varepsilon.$$

Similarly, if Q_i, $i \in M_2$, belong to $\{Q_j^{(2)}\}$ but not to $\{Q_j^{(1)}\}$, we find

$$\sum_{i \in M_2} h_w^{N-\alpha}(Q_i) \leq \sum_{j \in N_2} h_w^{N-\alpha}(Q_j^{(m)}) + 2^{-2}\varepsilon,$$

where N_1 and N_2 are disjoint. If we repeat this argument m times and add the inequalities thus obtained, it follows that

$$\sum_{k=1}^{m} \sum_{i \in M_k} h_w^{N-\alpha}(Q_i) \leq \sum_{k=1}^{m} \sum_{j \in N_k} h_w^{N-\alpha}(Q_j^{(m)}) + \varepsilon \sum_{k=1}^{m} 2^{-k}$$
$$< \widetilde{\mathcal{H}}_{w,\rho}^{N-\alpha}(E_m) + 2\varepsilon.$$

Finally, passing to the limit in the last inequality, we get

$$\widetilde{\mathcal{H}}_{w,\rho}^{N-\alpha}(E) \leq \sum_{i} h_w^{N-\alpha}(Q_i) = \sum_{k=1}^{\infty} \sum_{i \in M_k} h_w^{N-\alpha}(Q_i)$$
$$\leq \lim_{m \to \infty} \widetilde{\mathcal{H}}_{w,\rho}^{N-\alpha}(E_m) + 2\varepsilon,$$

and thus, $\widetilde{\mathcal{H}}_{w,\rho}^{N-\alpha}(E) \leq \lim_{m \to \infty} \widetilde{\mathcal{H}}_{w,\rho}^{N-\alpha}(E_m) = l$. $\square$

Proposition 3.4.23. *Let* $0 \leq \alpha < N$ *and* $0 < \rho \leq \infty$. *Let* w *be a weight. If* $K_1 \supset K_2 \supset \ldots$ *is a decreasing sequence of compact subsets of* $\mathbf{R}^N$, *then*

$$\widetilde{\mathcal{H}}_{w,\rho}^{N-\alpha}\left(\bigcap_{n=1}^{\infty} K_n\right) = \lim_{n \to \infty} \widetilde{\mathcal{H}}_{w,\rho}^{N-\alpha}(K_n). \qquad (3.4.23)$$

Proof. If we set $K = \bigcap_{n=1}^{\infty} K_n$, then obviously

$$\widetilde{\mathcal{H}}_{w,\rho}^{N-\alpha}(K) \leq \lim_{n \to \infty} \widetilde{\mathcal{H}}_{w,\rho}^{N-\alpha}(K_n);$$

we will prove the reverse inequality. Let $\varepsilon > 0$ be arbitrary, and choose a proper, dyadic covering $\{Q_j\}$ of K such that each Q_j has side-length $\leq \rho$ and

$$\sum_{j} h_w^{N-\alpha}(Q_j) < \widetilde{\mathcal{H}}_{w,\rho}^{N-\alpha}(K) + \varepsilon.$$

It is easy to see that $K_m \subset (\bigcup_j Q_j)^\circ$, if m is large enough. Thus, for large m,

$$\lim_{n \to \infty} \widetilde{\mathcal{H}}_{w,\rho}^{N-\alpha}(K_n) \leq \widetilde{\mathcal{H}}_{w,\rho}^{N-\alpha}(K_m) \leq \sum_{j} h_w^{N-\alpha}(Q_j) < \widetilde{\mathcal{H}}_{w,\rho}^{N-\alpha}(K) + \varepsilon,$$

and this proves the proposition. $\square$

Remark 3.4.24. It is readily seen that Proposition 3.4.23 remains true with $\widetilde{\mathcal{H}}_{w,\rho}^{N-\alpha}$ replaced by $\mathcal{H}_{w,\rho}^{N-\alpha}$. Indeed, the same proof works.

Having proved Proposition 3.4.22 and Proposition 3.4.23, the capacitability theorem below for the dyadic Hausdorff capacity $\widetilde{\mathcal{H}}_{w,\rho}^{N-\alpha}$ is an immediate consequence of Choquet's theorem, Theorem 3.2.8.

Theorem 3.4.25. Let $0 \le \alpha < N$ and $0 < \rho \le \infty$. Let w be a doubling weight. In the case $\rho = \infty$, suppose that w satisfies (3.4.21) above. Then, if E is a Suslin set,

$$\widetilde{\mathcal{H}}_{w,\rho}^{N-\alpha}(E) = \sup\{\widetilde{\mathcal{H}}_{w,\rho}^{N-\alpha}(K) \,;\, K \subset E,\, K \text{ compact}\}$$
$$= \inf\{\widetilde{\mathcal{H}}_{w,\rho}^{N-\alpha}(G) \,;\, E \subset G,\, G \text{ open}\}.$$

3.4.6. Frostman's lemma

We shall finally prove a weighted version of Frostman's lemma [42],[†] which gives a dual definition of $\mathcal{H}_{w,\rho}^{N-\alpha}$. Our proof is a modification of the standard proof in the non-weighted, homogeneous case (see L. Carleson [19, p. 7]).[‡]

First of all, we define a fractional maximal function.

Definition 3.4.26. Let $0 \le \alpha < N$ and $0 < \rho \le \infty$, and let w be a weight. The *weighted, fractional maximal function* $M_{w,\rho}^{N-\alpha}\mu$ of a Borel measure μ on $\mathbf{R}^N$ is defined by

$$M_{w,\rho}^{N-\alpha}\mu(x) = \sup_{0 < r \le \rho} \frac{|\mu|(B_r(x))}{h_w^{N-\alpha}(B_r(x))}, \quad x \in \mathbf{R}^N.$$

Theorem 3.4.27. Let $0 \le \alpha < N$ and $0 < \rho \le \infty$, and let w be a doubling weight. In the case $\rho = \infty$, suppose that w satisfies (3.4.21) above. Then there exists a positive constant C, depending only on α, N, and the doubling constant of w, such that if $E \subset \mathbf{R}^N$ is a Suslin set, then

$$C\mathcal{H}_{w,\rho}^{N-\alpha}(E) \le \sup\{\mu(E) \,;\, \mu \in \mathcal{M}^+(E),\, \|M_{w,\rho}^{N-\alpha}\mu\|_{L^\infty(\mathbf{R}^N)} \le 1\}$$
$$\le \mathcal{H}_{w,\rho}^{N-\alpha}(E). \tag{3.4.24}$$

Proof. The second inequality in (3.4.24) is obvious. We will prove the opposite inequality in the slightly more complicated case $0 < \rho < \infty$. By Theorem 3.4.25, we may assume that E is compact. For $n \in \mathbf{Z}$, let $\mathcal{Q}_n$ be the mesh of dyadic cubes with side-lengths 2^{-n}. Then choose $Q_0 \in \mathcal{Q}_m$ such that $E \subset Q_0$ and $2^{-m} \ge \rho$. Choose $k \in \mathbf{Z}$ so that $2^{-k} \le \rho < 2^{-k+1}$.

[†]A similar result is stated without a proof in Vodop'yanov [104, p. 207].

[‡]The theorem in the case $w = 1$ and $0 < \rho < \infty$ is formulated, but not proved in Carlsson–Maz'ya [22, p. 4].

We begin by defining μ_n, $n = k+1, k+2, \ldots$, as the measure with mass $h_w^{N-\alpha}(Q)$ and constant density on every cube $Q \in \mathcal{Q}_n$ that intersects E, and then modify μ_n as follows. If $\mu_n(Q) > h_w^{N-\alpha}(Q)$ for some $Q \in \mathcal{Q}_{n-1}$, we reduce the mass of μ_n on Q until it equals $h_w^{N-\alpha}(Q)$ (without increasing the mass on any of the subcubes of Q that belong to $\mathcal{Q}_n$). Otherwise, we leave μ_n unchanged. After a possible modification, we obtain the measure $\mu_{n,1}$. Next, if $n \geq k+2$, we modify $\mu_{n,1}$ in the same way with respect to the cubes $Q \in \mathcal{Q}_{n-2}$ and get the measure $\mu_{n,2}$. Continuing in this way, we obtain the measure $\mu_{n,n-k}$ after $n - k$ modifications. Let us denote $\mu_{n,n-k}$ again by μ_n. By the construction, we have $\mu_n(Q) \leq h_w^{N-\alpha}(Q)$ for every $Q \in \mathcal{Q}_l$, $k \leq l \leq n$. If $Q_0 = \bigcup_{j=1}^{2^{(k-m)N}} Q_j$, where $Q_j \in \mathcal{Q}_k$, it follows that

$$\mu_n(Q_0) = \sum_{j=1}^{2^{(k-m)N}} \mu_n(Q_j) \leq \sum_{j=1}^{2^{(k-m)N}} h_w^{N-\alpha}(Q_j) = 2^{(k-m)\alpha} h_w^{N-\alpha}(Q_0)$$

for $n = k+1, k+2, \ldots$. By weak compactness, we may assume that μ_n converges weakly to a measure $\mu \in \mathcal{M}^+(E)$.

We will now show that $\mu(E) \geq C\mathcal{H}_{w,\rho}^{N-\alpha}(E)$. Let $n \geq k+1$ be fixed. Note that every $x \in E$ belongs to one or several cubes Q_x such that $\mu_n(Q_x) = h_w^{N-\alpha}(Q_x)$. The maximal cubes Q_x yield a disjoint covering $\{Q_j\}$ of E satisfying

$$\mu_n(Q_0) = \sum_j \mu_n(Q_j) = \sum_j h_w^{N-\alpha}(Q_j).$$

If we for each j let B_j be the ball with the same center as Q_j and radius $\sqrt{N}$ times the side-length of Q_j, then $h_w^{N-\alpha}(Q_j) \geq C h_w^{N-\alpha}(B_j)$ since w is doubling, and we find

$$\mu_n(Q_0) \geq C \sum_j h_w^{N-\alpha}(B_j) \geq C\mathcal{H}_{w,\rho\sqrt{N}}^{N-\alpha}(E) \geq C\mathcal{H}_{w,\rho}^{N-\alpha}(E),$$

where the last inequality follows from Lemma 3.4.15(a). Using standard properties of weak convergence, we see that

$$\mu(E) = \mu(Q_0) \geq \limsup_{n \to \infty} \mu_n(Q_0) \geq \mathcal{H}_{w,\rho}^{N-\alpha}(E).$$

It remains to estimate $\|M_{w,\rho}^{N-\alpha}\mu\|_{L^\infty(\mathbf{R}^N)}$. Let B be a ball with radius r, where $2^{-l+1} < r \leq 2^{-l} \leq \rho$. If $\chi \in C_0^\infty(2B)$ and $\chi = 1$ on $\bar{B}$ and $0 \leq \chi \leq 1$ otherwise, then

$$\mu(B) \leq \int_{2B} \chi \, d\mu = \lim_{n \to \infty} \int_{2B} \chi \, d\mu_n.$$

Moreover, $\int_{2B} \chi \, d\mu_n \leq \mu_n(2B)$ for every n. Now, the ball $2B$ can be covered by C_N cubes $Q_j \in \mathcal{Q}_{l+1}$, and we obtain

$$\mu_n(2B) \leq \sum_{j=1}^{C_N} \mu_n(Q_j) \leq \sum_{j=1}^{C_N} h_w^{N-\alpha}(Q_j) \leq C h_w^{N-\alpha}(B).$$

This implies that $\mu(B) \leq C h_w^{N-\alpha}(B)$, and hence that $\|M_{w,\rho}^{N-\alpha}\mu\|_{L^\infty(\mathbf{R}^N)} \leq C$. The desired measure is therefore μ/C. $\square$

3.5. Variational capacities

In the present section, we have collected some results concerning variational capacities. We first show that the Bessel capacity $B_{m,p}^w$ is equivalent to the variational capacity $c_{m,p}^w$, defined below, on compact subsets of $\mathbf{R}^N$, if w is an A_p weight. Similar results relating the Hausdorff capacities $\mathcal{H}_{w,1}^{N-m}$ and $\mathcal{H}_{w,\infty}^{N-m}$ to the variational capacities $C_{m,1}^w$ and $\mathrm{Cap}_{m,1}^w$, respectively, are then established for A_1 weights w. The variational capacities $c_{m,p}^w$ and $C_{m,1}^w$ are naturally associated with the norms in the spaces $W_w^{m,p}(\mathbf{R}^N)$ and $W_w^{m,1}(\mathbf{R}^N)$. Hence, these results motivate the use of the Bessel capacity and the Hausdorff capacity in later sections for measuring exceptional sets for Sobolev functions. A consequence of the equivalence between $\mathcal{H}_{w,\infty}^{N-m}$ and $\mathrm{Cap}_{m,1}^w$ and Theorem 3.4.6 is a formula for the $\mathrm{Cap}_{m,1}^w$ capacity of a ball. Since the proof of this equivalence is quite deep and involves both an embedding theorem from Section 2.6 (and thus ultimately the boxing inequality in Section 2.4) and Frostman's lemma, the author has preferred to include a direct proof of the formula for the capacity of a ball. With the aid of this result, we finally show in Theorem 3.5.7 that a necessary and sufficient condition for the inequality

$$\left(\int_{\mathbf{R}^N} |u|^q \, d\mu \right)^{1/q} \leq C \int_{\mathbf{R}^N} |\nabla u| w \, dx \tag{3.5.1}$$

to hold for every function $u \in C^\infty(\mathbf{R}^N)$ is that

$$\sup_{a \in \mathbf{R}^N, \, r>0} \frac{\mu(B_r(a))^{1/q}}{\mathrm{Cap}_{1,1}^w(B_r(a))} < \infty.$$

3.5.1. The case $1 < p < \infty$

We begin by defining the variational capacity $c_{m,p}^w$.

Definition 3.5.1. Let m be an integer, $1 \leq m < N$, and $1 < p < \infty$, and let w be a weight. If $K \subset \mathbf{R}^N$ is compact, we define the *weighted, m-th order Sobolev capacity* $c_{m,p}^w(K)$ by

$$c_{m,p}^w(K) = \inf\{ \|\varphi\|_{W_w^{m,p}(\mathbf{R}^N)}^p \; ; \varphi \in C_0^\infty(\mathbf{R}^N) \text{ and } \varphi \geq 1 \text{ on } K\}.$$

We now show that the capacities $B_{m,p}^w$ and $c_{m,p}^w$ are equivalent on compact subsets of $\mathbf{R}^N$.

Theorem 3.5.2. *Let m be an integer, $1 \leq m < N$, and let $1 < p < \infty$. Let w be an A_p weight. Then there are two positive constants, C_1 and C_2, such that*

$$C_1 c^w_{m,p}(K) \leq B^w_{m,p}(K) \leq C_2 c^w_{m,p}(K) \tag{3.5.2}$$

for every compact set $K \subset \mathbf{R}^N$.

Our proof is modeled after a proof by Yu. G. Reshetnyak [94, pp. 840–841] and uses the lemma below.

Lemma 3.5.3. *Let $0 < \alpha < N$ and $1 < p < \infty$. Let w be an A_p weight. Then*

$$B^w_{\alpha,p}(K) = \inf\{\|f\|^p_{L^p_w(\mathbf{R}^N)} \; ; f \geq 0, \mathcal{G}_\alpha f \in C^\infty(\mathbf{R}^N), \text{ and}$$
$$\mathcal{G}_\alpha f(x) \geq 1 \text{ for every } x \in K\} \tag{3.5.3}$$

for every compact set $K \subset \mathbf{R}^N$.

Proof. If we temporarily denote the right-hand side in (3.5.3) by I, it suffices to show that $I \leq B^w_{\alpha,p}(K)$. Let $\varepsilon > 0$ be arbitrary, and let $f \geq 0$ satisfy $\mathcal{G}_\alpha f \geq 1$ on K and $\|f\|^p_{L^p_w(\mathbf{R}^N)} < B^w_{\alpha,p}(K) + \varepsilon$. We may obviously assume that $\inf_{x \in K} \mathcal{G}_\alpha f(x) = 1$. For $h > 0$, we define

$$K_h = \{x \in \mathbf{R}^N \; ; \text{dist}(x, K) \leq h\},$$

and let

$$\inf_{x \in K_h} \mathcal{G}_\alpha f(x) = 1 - \delta(h).$$

We claim that $\delta(h) \to 0$, as $h \to 0$. In fact, the set

$$G_s = \{x \in \mathbf{R}^N \; ; \mathcal{G}_\alpha f(x) > 1 - s\}, \quad 0 < s < 1,$$

is open since $\mathcal{G}_\alpha f$ is lower semicontinuous (this follows from Fatou's lemma). Because the distance between K and G_s^c is positive, G_s contains K_h, if h is small enough. It follows that $1 - \delta(h) \geq 1 - s$, so that $0 \leq \delta(h) \leq s$ for small h. This proves the claim. Let φ_h be the mollifier in Theorem 2.1.4, and let $f_h = (1 - \delta(h))^{-1} \varphi_h * f$. Then

$$\mathcal{G}_\alpha f_h(x) = (1 - \delta(h))^{-1} \varphi_h * \mathcal{G}_\alpha f(x) \geq 1$$

for $x \in K$, and $\mathcal{G}_\alpha f_h \in C^\infty(\mathbf{R}^N)$ because $\mathcal{G}_\alpha f$ is locally integrable on $\mathbf{R}^N$. We also know that $f_h \to f$ in $L^p_w(\mathbf{R}^N)$, which implies

$$I \leq \|f_h\|^p_{L^p_w(\mathbf{R}^N)} \leq \|f\|^p_{L^p_w(\mathbf{R}^N)} + \varepsilon < B^w_{\alpha,p}(K) + 2\varepsilon,$$

if h is small enough. $\square$

Proof of Theorem 3.5.2. Let $\varepsilon > 0$ be arbitrary, and let $\varphi \in C^\infty_0(\mathbf{R}^N)$

satisfy $\varphi \geq 1$ on K and $\|\varphi\|^p_{W^{m,p}_w(\mathbf{R}^N)} < c^w_{m,p}(K) + \varepsilon$. By Theorem 3.3.2, we have $\varphi = \mathcal{G}_m f$ for some $f \in L^p_w(\mathbf{R}^N)$ (more precisely, $f \in \mathcal{S}$) such that $\|f\|_{L^p_w(\mathbf{R}^N)} \leq C\|\varphi\|_{W^{m,p}_w(\mathbf{R}^N)}$. It follows that

$$1 \leq \varphi(x) = \mathcal{G}_m f(x) \leq \mathcal{G}_m |f|(x)$$

for $x \in K$, and

$$B^w_{m,p}(K) \leq \|f\|^p_{L^p_w(\mathbf{R}^N)} \leq C\|\varphi\|^p_{W^{m,p}_w(\mathbf{R}^N)} < C(c^w_{m,p}(K) + \varepsilon).$$

This proves the second inequality in (3.5.2).

Let $\psi \in C^\infty_0(\mathbf{R}^N)$ be such that $\operatorname{supp}\psi \subset B_2(0)$ and $\psi = 1$ on $\overline{B_1(0)}$. Suppose that $K \subset \overline{B_R(a)}$, where $R \geq 1$. We then set $\Psi(x) = \psi((x - a)/R)$, $x \in \mathbf{R}^N$. For an arbitrary $\varepsilon > 0$, we let $f \geq 0$ be such that $\mathcal{G}_m f \in C^\infty(\mathbf{R}^N)$, $\mathcal{G}_m f \geq 1$ on K, and $\|f\|^p_{L^p_w(\mathbf{R}^N)} < B^w_{m,p}(K) + \varepsilon$. Then $\varphi = \Psi\,\mathcal{G}_m f$ belongs to $C^\infty_0(\mathbf{R}^N)$ and $\varphi \geq 1$ on K. Furthermore, by the Leibniz rule and Theorem 3.3.2,

$$c^w_{m,p}(K) \leq \|\varphi\|^p_{W^{m,p}_w(\mathbf{R}^N)} \leq C\|\mathcal{G}_m f\|^p_{W^{m,p}_w(\mathbf{R}^N)}$$
$$\leq C\|f\|^p_{L^p_w(\mathbf{R}^N)} < C(B^w_{\alpha,p}(K) + \varepsilon),$$

and the theorem follows. $\quad\square$

3.5.2. The case $p = 1$

Definition 3.5.4. Let m be an integer, $1 \leq m < N$, and let w be a weight. If $E \subset \mathbf{R}^N$, we define the *weighted, m-th order variational capacity* $\operatorname{Cap}^w_{m,1}(E)$ by

$$\operatorname{Cap}^w_{m,1}(E) = \inf\left\{ \int_{\mathbf{R}^N} |\nabla^m \varphi| w \, dx \; ; \varphi \in C^\infty_0(\mathbf{R}^N), 0 \leq \varphi \leq 1, \text{ and } \right.$$
$$\left. \varphi = 1 \text{ in a neighbourhood of } E \right\}.$$

We similarly define the *weighted, m-th order Sobolev capacity* $\mathrm{C}^w_{m,1}(E)$ by

$$\mathrm{C}^w_{m,1}(E) = \inf\{\|\varphi\|_{W^{m,1}_w(\mathbf{R}^N)} \; ; \varphi \in C^\infty_0(\mathbf{R}^N), 0 \leq \varphi \leq 1, \text{ and }$$
$$\varphi = 1 \text{ in a neighbourhood of } E\}.$$

Theorem 3.5.5 below shows that $\mathcal{H}^{N-m}_{w,\infty}$ and $\mathcal{H}^{N-m}_{w,1}$ are equivalent to $\operatorname{Cap}^w_{m,1}$ and $\mathrm{C}^w_{m,1}$, respectively, where m is an integer, $1 \leq m < N$, at least when $w \in A_1$ and on compact subsets of $\mathbf{R}^N$. The corresponding results in the non-weighted case are due to D. R. Adams [4, p. 121] and A. Carlsson [22, p. 4].

Theorem 3.5.5. *Let m be an integer, $1 \leq m < N$, and let $w \in A_1$. Then there exist positive constants $C_1, C_2, D_1,$ and D_2, depending only on m, N, and the A_1 constant of w, such that, for every compact $K \subset \mathbf{R}^N$,*

$$C_1 \mathcal{H}^{N-m}_{w,\infty}(K) \leq \operatorname{Cap}^w_{m,1}(K) \leq C_2 \mathcal{H}^{N-m}_{w,\infty}(K) \qquad (3.5.4)$$

and

$$D_1 \mathcal{H}_{w,1}^{N-m}(K) \leq C_{m,1}^w(K) \leq D_2 \mathcal{H}_{w,1}^{N-m}(K). \tag{3.5.5}$$

Proof. We will only prove the equivalence between $\mathcal{H}_{w,1}^{N-m}$ and $C_{m,1}^w$. The proof of the equivalence between $\mathcal{H}_{w,\infty}^{N-m}$ and $\text{Cap}_{m,1}^w$ is completely analogous. Let $K \subset \mathbf{R}^N$ be compact. It follows from Frostman's lemma (Theorem 3.4.27) that there exists a measure $\mu \in \mathcal{M}^+(K)$, satisfying $\|M_{w,1}^{N-m}\mu\|_{L^\infty(\mathbf{R}^N)} \leq 1$ and $\mathcal{H}_{w,1}^{N-m}(K) \leq C\mu(K)$. Let $\varphi \in C_0^\infty(\mathbf{R}^N)$ be such that $0 \leq \varphi \leq 1$ and $\varphi = 1$ in a neighbourhood of K. An application of Theorem 2.6.4 shows that

$$\mathcal{H}_{w,1}^{N-m}(K) \leq C \int_{\mathbf{R}^N} \varphi \, d\mu \leq C\|\varphi\|_{W_w^{m,1}(\mathbf{R}^N)},$$

and if we take the infimum over all such φ, we get the first inequality in (3.5.5).

For the proof of the second inequality, let $\varepsilon > 0$ be arbitrary, and then cover K with balls B_j, $j = 1, \ldots, s$, such that each B_j has radius $r_j \leq 1$ and

$$\sum_{j=1}^s h_w^{N-\alpha}(B_j) < \mathcal{H}_{w,1}^{N-m}(K) + \varepsilon.$$

It is well-known (see Harvey–Polking [51, p. 43]) that one can find functions $\varphi_j \in C_0^\infty(\mathbf{R}^N)$, $j = 1, \ldots, s$, supported in $2B_j$, such that $0 \leq \varphi \leq 1$, $|D^\alpha \varphi_j| \leq Cr_j^{-|\alpha|}$ for every α and $\varphi = \sum_{j=1}^s \varphi_j = 1$ on $\bigcup_{j=1}^s B_j$. This implies that

$$C_{m,1}^w(K) \leq \|\varphi\|_{W_w^{m,1}(\mathbf{R}^N)} \leq \sum_{k=1}^m \sum_{j=1}^s \int_{2B_j} |\nabla^k \varphi_j| w \, dx$$

$$\leq \sum_{k=1}^m \sum_{j=1}^s \frac{w(2B_j)}{r_j^k} \leq C \sum_{j=1}^s \frac{w(2B_j)}{r_j^m}$$

$$\leq C \sum_{j=1}^s \frac{w(B_j)}{r_j^m} < C(\mathcal{H}_{w,1}^{N-m}(K) + \varepsilon).$$

This proves the second inequality in (3.5.5) since ε was arbitrary. $\square$

Theorem 3.5.6. *Let $w \in A_1$, and let $m \geq 1$ be an integer. Then there are positive constants C_1 and C_2, which only depend on m, N, and the A_1 constant of w, such that, for every ball $B_r(a)$,*

$$C_1 \inf_{t \geq r} \frac{w(B_t(a))}{t^\alpha} \leq \text{Cap}_{m,1}^w(B_r(a)) \leq C_2 \inf_{t \geq r} \frac{w(B_t(a))}{t^\alpha}. \tag{3.5.6}$$

Proof. Let us temporarily use the notation

$$h_w(B_r(a)) = \inf_{t \geq r} \frac{w(B_t(a))}{t^\alpha}.$$

We will first prove the second inequality in (3.5.6). To this end, let $t \geq 4r$ be arbitrary, and let $\varphi \in C_0^\infty(\mathbf{R}^N)$, with support in $B_2(0)$, be such that $\varphi = 1$ on $\overline{B_1(0)}$. If we set $\psi(x) = \varphi(2(x - a)/t)$, $x \in \mathbf{R}^N$, then $\operatorname{supp} \psi \subset B_t(a)$ and $\psi = 1$ on $\overline{B_{t/2}(a)} \supset B_r(a)$. We find

$$\operatorname{Cap}_{m,1}^w(B_r(a)) \leq \int_{\mathbf{R}^N} |\nabla^m \psi| w \, dx \leq C \frac{w(B_t(a))}{t^\alpha}.$$

Since $t \geq 4r$ was arbitrary, it follows that $\operatorname{Cap}_{m,1}^w(B_r(a)) \leq h_w(B_{4r}(a))$. But if $r \leq t \leq 4r$, then, by the doubling property of w,

$$\frac{w(B_t(a))}{t^\alpha} \geq C \frac{w(B_{4r}(a))}{(4r)^\alpha},$$

and hence, $h_w(B_r(a)) \geq C h_w(B_{4r}(a))$. This proves the second inequality in (3.5.6).

We now prove the first inequality in (3.5.6). Let $\varepsilon > 0$ be arbitrary, and let $\varphi \in C_0^\infty(\mathbf{R}^N)$ be such that $\varphi = 1$ in a neighbourhood of $B_r(a)$ and

$$\int_{\mathbf{R}^N} |\nabla^m \varphi| w \, dx < \operatorname{Cap}_{m,1}^w(B_r(a)) + \varepsilon. \tag{3.5.7}$$

Because $D^\alpha \varphi = 0$ on $B_r(a)$ for $|\alpha| = m$, we have

$$\begin{aligned}
1 = \varphi(a) &\leq C \int_{|x-a| \geq r} \frac{|\nabla^m \varphi(x)|}{|x - a|^{N-m}} \, dx \\
&= C \int_{|x-a| \geq r} \frac{1}{|x - a|^{N-m} w(x)} |\nabla^m \varphi(x)| w(x) \, dx \\
&\leq C \operatorname*{ess\,sup}_{|x-a| \geq r} \left(\frac{1}{|x - a|^{N-m} w(x)} \right) \int_{\mathbf{R}^N} |\nabla^m \varphi| w \, dx,
\end{aligned}$$

so by (3.5.7),

$$1 \leq C \operatorname*{ess\,sup}_{|x-a| \geq r} \left(\frac{1}{|x - a|^{N-m} w(x)} \right) \operatorname{Cap}_{m,1}^w(B_r(a)).$$

To get the lower bound for $\operatorname{Cap}_{m,1}^w(B_r(a))$, we thus have to show that

$$\operatorname*{ess\,sup}_{|x-a| \geq r} \left(\frac{1}{|x - a|^{N-m} w(x)} \right) \leq \frac{C}{h_w(B_r(a))}.$$

Let $E_n = \{ x \in \mathbf{R}^N \, ; \, nr \leq |x - a| < (n+1)r \}$, $n = 1, 2, \dots$. By the A_1 condition,

$$w(x) \geq \fint_{B_{(n+1)r}(a)} w \, dy$$

for a.e. $x \in E_n$. Hence, for such x,

$$\frac{1}{|x-a|^{N-m}w(x)} \leq C(nr)^{m-N}\frac{|B_{(n+1)r}(a)|}{w(B_{(n+1)r}(a))}$$

$$= C\left(\frac{n+1}{n}\right)^{N-m}\frac{((n+1)r)^m}{w(B_{(n+1)r}(a))}$$

$$\leq \frac{C}{h_w(B_r(a))}.$$

Since $\{x \in \mathbf{R}^N \,;\, |x-a| \geq r\} = \bigcup_{n=1}^{\infty} E_n$, we see that

$$\frac{1}{|x-a|^{N-m}w(x)} \leq \frac{C}{h_w(B_r(a))}$$

for a.e. x such that $|x-a| \geq r$, and we are done. $\square$

3.5.3. An embedding theorem

Using Theorem 2.6.1 and Theorem 3.5.6, we end this section by giving a necessary and sufficient condition for the inequality (3.5.1) in terms of capacities of balls.

Theorem 3.5.7. Let $w \in A_1$, and let $1 \leq q < \infty$. Suppose that μ is a positive Radon measure, satisfying

$$M = \sup_{a \in \mathbf{R}^N, \, r>0} \frac{\mu(B_r(a))^{1/q}}{\mathrm{Cap}_{1,1}^w(B_r(a))} < \infty.$$

Then the inequality

$$\left(\int_{\mathbf{R}^N} |u|^q \, d\mu\right)^{1/q} \leq C \int_{\mathbf{R}^N} |\nabla u| w \, dx \qquad (3.5.8)$$

holds for every $u \in C_0^{\infty}(\mathbf{R}^N)$ with $C = C'M$, where C' only depends on N, q, and the A_1 constant of w. Conversely, if there exists a constant C such that the inequality (3.5.8) holds for every $u \in C_0^{\infty}(\mathbf{R}^N)$, then $C \geq C'M$, with C' as before. In particular, M is finite.

Proof. According to Theorem 3.5.6, M is comparable to

$$M' = \sup_{a \in \mathbf{R}^N, \, r>0} \frac{\mu(B_r(a))^{1/q}}{\inf_{t \geq r} t^{-1}w(B_t(a))}.$$

Furthermore, it follows from Theorem 2.6.1 that it is sufficient to show that

$$M' = \sup_{a \in \mathbf{R}^N, \, r>0} \frac{r\mu(B_r(a))^{1/q}}{w(B_r(a))}.$$

But, by properties of the supremum,

$$
\begin{aligned}
M' &= \sup_{a \in \mathbf{R}^N} \sup_{r>0} \mu(B_r(a))^{1/q} \sup_{t \geq r} \frac{t}{w(B_t(a))} \\
&= \sup_{a \in \mathbf{R}^N} \sup_{r>0} \sup_{t \geq r} \mu(B_r(a))^{1/q} \frac{t}{w(B_t(a))} \\
&= \sup_{a \in \mathbf{R}^N} \sup_{t>0} \frac{t}{w(B_t(a))} \sup_{0<r \leq t} \mu(B_r(a))^{1/q} \\
&= \sup_{a \in \mathbf{R}^N} \sup_{t>0} \frac{t}{w(B_t(a))} \mu(B_t(a))^{1/q}. \quad \square
\end{aligned}
$$

3.6. Thinness: The case $1 < p < \infty$

The following definition of thinness was given by D. R. Adams in [5, p. 89], with the difference that Adams' definition involves the homogeneous Riesz capacity $R_{\alpha,p}^w$ instead of the inhomogeneous capacity $R_{\alpha,p;1}^w$.

Definition 3.6.1. Let $0 < \alpha < N$ and $1 < p < \infty$, and let w be a weight. We say that a set $E \subset \mathbf{R}^N$ is $R_{\alpha,p;1}^w$-*thin* at a point $a \in \mathbf{R}^N$ if

$$
\int_0^1 \left(\frac{t^{\alpha p}}{w(B_t(a))} \right)^{p'-1} \frac{dt}{t} = \infty \tag{3.6.1}
$$

and

$$
\int_0^1 \left(\frac{t^{\alpha p} R_{\alpha,p;1}^w(E \cap B_t(a))}{w(B_t(a))} \right)^{p'-1} \frac{dt}{t} < \infty. \tag{3.6.2}
$$

If E is not $R_{\alpha,p;1}^w$-thin at a, then E is said to be $R_{\alpha,p;1}^w$-*thick* at a. The set of points where E is $R_{\alpha,p;1}^w$-thin is denoted $e_{\alpha,p;1}^w(E)$ and its complement with respect to $\mathbf{R}^N$ is the set $b_{\alpha,p;1}^w(E)$.

Remark 3.6.2. Note that the first condition in the definition is always fulfilled in the non-weighted case, assuming that $\alpha p \leq N$.

In view of Theorem 3.3.9, this definition is equivalent to Adams' definition as soon as $R_{\alpha,p}^w$ is non-trivial. With $\alpha = 1$ and $p = 2$, the definition coincides with the criterion for regularity of a point for the Dirichlet problem for a second order, degenerate, elliptic partial differential operator in divergence form obtained earlier by E. B. Fabes, D. S. Jerison, and C. E. Kenig [35]. The corresponding definition in the non-weighted case is due to N. G. Meyers [78].

We intend here to show that this definition of thinness has the Kellogg property.[†]

[†]For Adams' definition, this has been obtained by S. K. Vodop'yanov [104] [105].

Theorem 3.6.3. *Let $0 < \alpha < N$ and $1 < p < \infty$, and let w be an A_p weight. Then*

$$R^w_{\alpha,p;1}(E \cap e^w_{\alpha,p;1}(E)) = 0 \qquad (3.6.3)$$

for every set $E \subset \mathbf{R}^N$.

With $w = 1$, this theorem is due to L. I. Hedberg [54] for $2 - \alpha/N < p < N/\alpha$ and to Hedberg and Th. H. Wolff [58] in the general case.[†] In Section 3.6.4, we investigate another definition of thinness that is equivalent to the definition above in the non-weighted case.

3.6.1. Preliminary considerations

The proof of Theorem 3.6.3 requires considerable preparation. We shall first introduce a new capacity, which we later show is equivalent to the inhomogeneous Riesz capacity $R^w_{\alpha,p;1}$ in Section 3.6.2.

Let $0 < \alpha < N$ and $1 < p < \infty$, and let w be an A_p weight. Let χ be the characteristic function for the unit ball $B_1(0)$. For $r > 0$, $x \in \mathbf{R}^N$, and $(y,t) \in \mathbf{R}^N \times \mathbf{R} \cong \mathbf{R}^{N+1}$, define the kernel k on $\mathbf{R}^N \times \mathbf{R}^{N+1}$ by

$$k(x,(y,t)) = \begin{cases} t^{-(N-\alpha)}\chi((x-y)/t), & \text{if } 0 < t < 1, \\ 0, & \text{otherwise.} \end{cases}$$

We also define a measure ν on $\mathbf{R}^{N+1}$ by

$$d\nu(y,t) = \begin{cases} w'(y)\,dy\,\dfrac{dt}{t}, & \text{if } 0 < t < 1, \\ 0, & \text{otherwise,} \end{cases}$$

where, as before, $w'(y) = w(y)^{-1/(p-1)}$. If f is a function on $\mathbf{R}^{N+1}$, then

$$k(x, f\nu) = \int_0^1 \left(\int_{|x-y|<t} \frac{f(y,t)}{t^{N-\alpha}} w'(y)\,dy \right) \frac{dt}{t},$$

and if $\mu \in \mathcal{M}(\mathbf{R}^N)$, then

$$k(\mu,(y,t)) = \begin{cases} t^{-(N-\alpha)}\mu(B_t(y)), & \text{if } 0 < t < 1, \\ 0, & \text{otherwise.} \end{cases}$$

We will denote by $\mathcal{R}^w_{\alpha,p}$ the L^p-capacity associated with the kernel k and the measure ν, that is,

$$\mathcal{R}^w_{\alpha,p}(E) = \{\|f\|^p_{L^p_\nu(\mathbf{R}^{N+1})} \,;\, f \in L^p_\nu(\mathbf{R}^{N+1})^+, \, k(x, f\nu) \geq 1 \text{ for every } x \in E\}.$$

[†] For a detailed account of the history of the subject of this section as well as further references, see Adams–Hedberg [7, pp. 190–191].

The nonlinear potential $\mathcal{V}^\mu = V^\mu_{k,\nu,p}$ of $\mu \in \mathcal{M}^+(\mathbf{R}^N)$ is given by

$$\mathcal{V}^\mu(x) = \int_0^1 \left(\int_{|x-y|<t} \left(\frac{\mu(B_t(y))}{t^{N-\alpha}} \right)^{p'-1} \frac{w'(y)}{t^{N-\alpha}} \, dy \right) \frac{dt}{t}, \quad x \in \mathbf{R}^N. \qquad (3.6.4)$$

We now establish the existence of capacitary measures and capacitary potentials for general sets with respect to the capacity $\mathcal{R}^w_{\alpha,p}$.

Proposition 3.6.4. *Let $0 < \alpha < N$ and $1 < p < \infty$, and let w be an A_p weight. If E is an arbitrary subset of $\mathbf{R}^N$, then there exists a measure $\mu^E \in \mathcal{M}^+(\overline{E})$ such that*

$$\mathcal{V}^{\mu^E}(x) \geq 1 \quad \text{for } \mathcal{R}^w_{\alpha,p}\text{-q.e. } x \in E,$$

$$\mathcal{V}^{\mu^E}(x) \leq 1 \quad \text{for every } x \in \operatorname{supp} \mu^E,$$

and

$$\mu^E(\overline{E}) = \mathcal{R}^w_{\alpha,p}(E).$$

To prove this result, it suffices, according to Proposition 3.2.15, to find a dense subset D of $L^p_\nu(\mathbf{R}^{N+1})$ such that if $\varphi \in D$, then the potential $k(\,\cdot\,, \varphi \nu)$ is continuous on $\mathbf{R}^N$ and, furthermore, $\lim_{|x|\to\infty} k(x, \varphi \nu) = 0$. Let D be the set of functions φ of the form $\varphi = t^{1/p} \psi / w'$, where $\psi \in C_0(\mathbf{R}^{N+1})$. Note that if $\varphi \in D$, then the function

$$k(x, \varphi \nu) = \int_0^1 \left(\int_{|x-y|<t} \psi(y,t) \, dy \right) t^{1/p-(N-\alpha)} \frac{dt}{t}$$

is continuous on $\mathbf{R}^N$. Moreover, $k(\,\cdot\,, \varphi \nu)$ has compact support.

Lemma 3.6.5. *The set D, as defined above, is dense in $L^p_\nu(\mathbf{R}^{N+1})$.*

Proof. Let $f \in L^p_\nu(\mathbf{R}^{N+1})$. If we define $g = t^{-1/p} f w'$, then

$$\int_0^1 \left(\int_{\mathbf{R}^N} |g|^p w \, dy \right) dt = \int_0^1 \left(\int_{\mathbf{R}^N} |f|^p w' \, dy \right) \frac{dt}{t}. \qquad (3.6.5)$$

Now choose $\psi_n \in C_0(\mathbf{R}^{N+1})$ such that $\psi_n \to g$ in the norm given by the left-hand side of (3.6.5). Put $\varphi_n = t^{1/p} \psi_n / w'$. Then every φ_n belongs to D, and

$$\int_0^1 \left(\int_{\mathbf{R}^N} |f - \varphi_n|^p w' \, dy \right) \frac{dt}{t} = \int_0^1 \left(\int_{\mathbf{R}^N} |g - \psi_n|^p w \, dy \right) dt,$$

which proves the lemma. $\square$

If we for $0 < \rho \leq \infty$ and a measure $\mu \in \mathcal{M}^+(\mathbf{R}^N)$ define the potential W^μ_ρ by

$$W^\mu_\rho(x) = \int_0^\rho \left(\frac{t^{\alpha p} \mu(B_t(x))}{w(B_t(x))} \right)^{p'-1} \frac{dt}{t}, \quad x \in \mathbf{R}^N, \qquad (3.6.6)$$

then using the A_p condition and Remark 1.2.4.2, we easily see that there are constants C_1 and C_2 such that

$$C_1 W_{1/2}^\mu(x) \le \mathcal{V}^\mu(x) \le C_2 W_2^\mu(x). \tag{3.6.7}$$

for every $x \in \mathbf{R}^N$.

3.6.2. A Wolff type inequality

We next show that $\mathcal{R}_{\alpha,p}^w$ is equivalent to the Riesz capacity $R_{\alpha,p;1}^w$. In the same way as in the proof of Theorem 3.3.7, using the dual definitions, it is enough to prove that the energies $\int_{\mathbf{R}^N} V_1^\mu \, d\mu$ and $\int_{\mathbf{R}^N} \mathcal{V}^\mu \, d\mu$ are comparable for every measure $\mu \in \mathcal{M}^+(\mathbf{R}^N)$. To this end, we will show that

$$C_1 \int_{\mathbf{R}^N} V_\rho^\mu \, d\mu \le \int_{\mathbf{R}^N} W_\rho^\mu \, d\mu \le C_2 \int_{\mathbf{R}^N} V_\rho^\mu \, d\mu \tag{3.6.8}$$

for $0 < \rho \le \infty$. The second inequality in (3.6.8) is a consequence of the pointwise estimate

$$W_\rho^\mu(x) \le C V_{2\rho}^\mu(x), \tag{3.6.9}$$

which we establish below. With $w = 1$ and $\rho = \infty$, the reverse inequality holds for $2 - \alpha/N < p < \infty$ (see D. R. Adams [3, p. 909]), but is otherwise false, as can be seen by taking μ to be a Dirac measure, in which case V_∞^μ is identically infinite. With this background, it may be surprising that the inequality to the left in (3.6.8) still is true. In the non-weighted case and with $\rho = \infty$, this is a result by Th. H. Wolff [58, p. 165]. Later, D. R. Adams [5, p. 77] showed that Wolff's inequality, with weights, may be proved using Theorem 1.2.21. The proof presented below for general ρ uses the same ideas as Adams' proof, but is slightly more complicated in the case $0 < \rho < \infty$.

Theorem 3.6.6. *Let $0 < \alpha < N$, $1 < p < \infty$, and $0 < \rho \le \infty$. Let w be an A_p weight. Then there are two positive constants, C_1 and C_2, which only depend on α, N, p, ρ, and the A_p constant of w, such that*

$$C_1 \int_{\mathbf{R}^N} V_\rho^\mu \, d\mu \le \int_{\mathbf{R}^N} W_\rho^\mu \, d\mu \le C_2 \int_{\mathbf{R}^N} V_\rho^\mu \, d\mu \tag{3.6.10}$$

for every measure $\mu \in \mathcal{M}^+(\mathbf{R}^N)$.

Proof. We first consider the inequality to the left in (3.6.10). By the identity (3.3.2) and Corollary 3.1.4,

$$\int_{\mathbf{R}^N} V_\rho^\mu \, d\mu = \int_{\mathbf{R}^N} (\mathcal{I}_{\alpha,\rho}\mu)^{p'} w' \, dx \le C \int_{\mathbf{R}^N} (M_{\alpha,\rho/4}\mu)^{p'} w' \, dx.$$

Let the function $\mathcal{I}^{p'}_{\alpha,\rho/2}\mu$ be defined by

$$\mathcal{I}^{p'}_{\alpha,\rho/2}\mu(x) = \left(\int_0^{\rho/2} \left(\frac{\mu(B_t(x))}{t^{N-\alpha}} \right)^{p'} \frac{dt}{t} \right)^{1/p'}, \quad x \in \mathbf{R}^N.$$

We then have $M_{\alpha,\rho/4}\mu \leq C \mathcal{I}^{p'}_{\alpha,\rho/2}\mu$, since, if $0 < r < \frac{1}{4}\rho$,

$$\mathcal{I}^{p'}_{\alpha,\rho/2}\mu(x) \geq \left(\int_r^{2r} \left(\frac{\mu(B_t(x))}{t^{N-\alpha}} \right)^{p'} \frac{dt}{t} \right)^{1/p'} \geq \frac{(\log 2)^{1/p'}}{2^{N-\alpha}} \frac{\mu(B_r(x))}{r^{N-\alpha}}.$$

It follows that

$$\int_{\mathbf{R}^N} (M_{\alpha,\rho/4}\mu)^{p'} w' \, dx \leq C \int_{\mathbf{R}^N} (\mathcal{I}^{p'}_{\alpha,\rho/2}\mu)^{p'} w' \, dx$$

$$= C \int_0^{\rho/2} \left(\int_{\mathbf{R}^N} \mu(B_t(x))^{p'} w'(x) \, dx \right) t^{-(N-\alpha)p'} \frac{dt}{t}.$$

We now get the desired bound for the energy of μ, if we notice that

$$\int_{\mathbf{R}^N} \mu(B_t(x))^{p'} w'(x) \, dx = \int_{\mathbf{R}^N} \left(\int_{|x-y|<t} \mu(B_t(x))^{p'-1} \, d\mu(y) \right) w'(x) \, dx$$

$$\leq \int_{\mathbf{R}^N} \left(\int_{|x-y|<t} \mu(B_{2t}(y))^{p'-1} \, d\mu(y) \right) w'(x) \, dx$$

$$= \int_{\mathbf{R}^N} \mu(B_{2t}(y))^{p'-1} w'(B_t(y)) \, d\mu(y),$$

and hence, by the A_p condition and the doubling property of w,

$$\int_0^{\rho/2} \left(\int_{\mathbf{R}^N} \mu(B_t(x))^{p'} w'(x) \, dx \right) t^{-(N-\alpha)p'} \frac{dt}{t}$$

$$\leq \int_{\mathbf{R}^N} \left(\int_0^{\rho/2} \frac{\mu(B_{2t}(y))^{p'-1}}{t^{(N-\alpha)p'}} w'(B_t(y)) \frac{dt}{t} \right) d\mu(y)$$

$$\leq C \int_{\mathbf{R}^N} W_\rho^\mu(y) \, d\mu(y),$$

which proves the first inequality in (3.6.10).

The reverse inequality follows from the estimate (3.6.9), since

$$\int_{\mathbf{R}^N} V_{2\rho}^\mu \, d\mu = \int_{\mathbf{R}^N} (\mathcal{I}_{\alpha,2\rho}\mu)^{p'} w' \, dx \leq C \int_{\mathbf{R}^N} (\mathcal{I}_{\alpha,\rho}\mu)^{p'} \, dx = \int_{\mathbf{R}^N} V_\rho^\mu \, d\mu.$$

We now prove (3.6.9). If we let $f = (\mathcal{I}_{\alpha,2\rho}\mu)^{p'-1}$, then by (2.4.6),

$$V_{2\rho}^\mu(x) = \int_{|x-y|<2\rho} \frac{f(y)w'(y)}{|x-y|^{N-\alpha}} \, dy$$

$$= (N - \alpha) \int_0^{2\rho} \left(\int_{|x-y|<t} f(y)w'(y)\,dy \right) t^{-(N-\alpha)} \frac{dt}{t}$$

$$+ (2\rho)^{-(N-\alpha)} \int_{|x-y|<2\rho} f(y)w'(y)\,dy$$

$$\geq (N - \alpha) \int_0^{\rho} \left(\int_{|x-y|<t} f(y)w'(y)\,dy \right) t^{-(N-\alpha)} \frac{dt}{t}.$$

Further, if $|x - y| < t < \rho$, then $B_t(x) \subset B_{2t}(y)$, and we get

$$f(y) = \left(\int_{|y-z|<2\rho} \frac{d\mu(z)}{|y-z|^{N-\alpha}} \right)^{p'-1}$$

$$\geq \left(\int_{|y-z|<2t} \frac{d\mu(z)}{|y-z|^{N-\alpha}} \right)^{p'-1} \geq \left(\frac{\mu(B_t(x))}{(2t)^{N-\alpha}} \right)^{p'-1}.$$

We thus have

$$V_{2\rho}^{\mu}(x) \geq C \int_0^{\rho} \left(\int_{|x-y|<t} \left(\frac{\mu(B_t(x))}{t^{N-\alpha}} \right)^{p'-1} w'(y)\,dy \right) t^{-(N-\alpha)} \frac{dt}{t}$$

$$= C \int_0^{\rho} \frac{\mu(B_t(x))^{p'-1}}{t^{(N-\alpha)p'}} w'(B_t(x)) \frac{dt}{t} \geq C W_{\rho}^{\mu}(x),$$

where the last inequality follows from Remark 1.2.4.2. □

Corollary 3.6.7. *Let* $0 < \alpha < N$ *and* $1 < p < \infty$. *Let* w *be an* A_p *weight. Then there exist two positive constants* C_1 *and* C_2, *which only depend on* α, N, p, *and the* A_p *constant of* w, *such that, for every set* $E \subset \mathbf{R}^N$,

$$C_1 R_{\alpha,p;1}^w(E) \leq \mathcal{R}_{\alpha,p}^w(E) \leq C_1 R_{\alpha,p;1}^w(E).$$

Proof. Two applications each of Corollary 3.1.4, Theorem 3.6.6, and the inequalities in (3.6.7) shows that

$$\int_{\mathbf{R}^N} V_1^{\mu}\,d\mu \leq C \int_{\mathbf{R}^N} V_{1/2}^{\mu}\,d\mu \leq C \int_{\mathbf{R}^N} W_{1/2}^{\mu}\,d\mu \leq C \int_{\mathbf{R}^N} V_1^{\mu}\,d\mu$$

$$\leq C \int_{\mathbf{R}^N} W_2^{\mu}\,d\mu \leq C \int_{\mathbf{R}^N} V_2^{\mu}\,d\mu \leq C \int_{\mathbf{R}^N} V_1^{\mu}\,d\mu,$$

so that

$$C_1 \int_{\mathbf{R}^N} V_1^{\mu}\,d\mu \leq \int_{\mathbf{R}^N} \mathcal{V}_1^{\mu}\,d\mu \leq C_2 \int_{\mathbf{R}^N} V_1^{\mu}\,d\mu$$

for $\mu \in \mathcal{M}^+(\mathbf{R}^N)$. These inequalities now imply the statement of the corollary as in Theorem 3.3.7. □

3.6.3. Proof of the Kellogg property

For the proof of Theorem 3.6.3, we need several lemmas, the first of which is a weighted version of a result by D. R. Adams and N. G. Meyers [8, p. 190]; see also Adams–Hedberg [7, p. 175].

Lemma 3.6.8. *Let $0 < \alpha < N$ and $1 < p < \infty$, and let w be an A_p weight. Suppose that $E \subset \mathbf{R}^N$ is a Suslin set such that $0 < R^w_{\alpha,p;1}(E) < \infty$, and let μ^E be the capacitary measure of E. Then, for any open set $G \subset \mathbf{R}^N$,*

$$\mu^E(G) \le R^w_{\alpha,p;1}(G \cap E). \tag{3.6.11}$$

Proof. By Proposition 3.2.9, there is a sequence $\{K_n\}_{n=1}^\infty$ of compact subsets of E such that $\lim_{n\to\infty} R^w_{\alpha,p;1}(K_n) = R^w_{\alpha,p;1}(E)$. Furthermore, by the proof of Proposition 3.2.15, we may assume that the capacitary measures $\mu_n = \mu^{K_n}$ converge weakly to μ^E. Let $K \subset G$ be compact and set $\sigma_n = \mu_n|_K$. We then have $V_1^{\sigma_n} \le 1$ on $\operatorname{supp} \sigma_n \subset K \cap K_n$, so by Proposition 3.2.14,

$$\sigma_n(K) = \sigma_n(K \cap K_n) \le R^w_{\alpha,p;1}(K \cap K_n) \le R^w_{\alpha,p;1}(G \cap E),$$

which implies that

$$\mu_n(G) = \sup_{K \subset G} \sigma_n(K) \le R^w_{\alpha,p;1}(G \cap E).$$

Now let $\varphi \in C_0(G)$ satisfy $\varphi \le 1$ on G. Then

$$\int_G \varphi \, d\mu^E = \lim_{n\to\infty} \int_G \varphi \, d\mu_n \le \lim_{n\to\infty} \mu_n(G) \le R^w_{\alpha,p;1}(G \cap E).$$

But $\mu^E(G)$ equals the supremum of $\int_G \varphi \, d\mu^E$ over all such φ. This proves the inequality (3.6.11). $\square$

According to the next lemma, the $R^w_{\alpha,p;1}$-capacitary density of a set is 0 at every point, where the set is $R^w_{\alpha,p;1}$-thin. The proof utilizes Lemma 3.6.10 below, which shows that a set is $R^w_{\alpha,p;1}$-thick at all of its interior points.

Lemma 3.6.9. *Let $0 < \alpha < N$ and $1 < p < \infty$, and let w be an A_p weight. Let E be a subset of $\mathbf{R}^N$. Suppose that E is $R^w_{\alpha,p;1}$-thin at a point $a \in \mathbf{R}^N$. We then have*

$$\lim_{r\to 0} \frac{R^w_{\alpha,p;1}(E \cap B_r(a))}{R^w_{\alpha,p;1}(B_r(a))} = 0. \tag{3.6.12}$$

Lemma 3.6.10. *Let $0 < \alpha < N$ and $1 < p < \infty$, and let w be an A_p weight. Then, if (3.6.1) holds for some $a \in \mathbf{R}^N$, it follows that*

$$\int_0^1 \left(\frac{t^{\alpha p} R^w_{\alpha,p;1}(B_t(a))}{w(B_t(a))} \right)^{p'-1} \frac{dt}{t} = \infty.$$

Proof of Lemma 3.6.9. We introduce three functions,

$$\varphi(t) = R_{\alpha,p;1}^w(E \cap B_t(a))^{p'-1},$$

$$\psi(t) = \left(\frac{t^{\alpha p}}{w(B_t(a))}\right)^{p'-1}\frac{1}{t},$$

and

$$\Psi(t) = \int_t^2 \psi(r)\, dr,$$

all defined for $0 < t \le 2$. Notice that $\Psi(2) = 0$ and $\Psi'(t) = -\psi(t)$, so integration by parts yields,

$$\int_t^2 \varphi(r)\psi(r)\, dr = \varphi(t)\Psi(t) + \int_t^2 \Psi(r)\, d\varphi(r). \qquad (3.6.13)$$

Since φ and Ψ are nonnegative and φ is nondecreasing, it follows from (3.6.2) and (3.6.13) that

$$\int_0^2 \Psi(r)\, d\varphi(r) < \infty.$$

The identity (3.6.13) thus implies that the limit $l = \lim_{t\to 0} \varphi(t)\Psi(t)$ exists. If we can show that $l = 0$, then (3.6.12) will follow from Theorem 3.3.10. Suppose that $l > 0$. Using Theorem 3.3.10, we find

$$\left(\frac{R_{\alpha,p;1}^w(E \cap B_t(a))}{R_{\alpha,p;1}^w(B_t(a))}\right)^{p'-1} \ge C\varphi(t)\Psi(t) \ge \frac{Cl}{2},$$

that is,

$$R_{\alpha,p;1}^w(E \cap B_t(a))^{p'-1} \ge \frac{Cl}{2} R_{\alpha,p;1}^w(B_t(a))^{p'-1},$$

if t is small enough, $0 < t \le t_0 \le 1$, say. This contradicts (3.6.2), since

$$\int_0^1 \left(\frac{r^{\alpha p}R_{\alpha,p;1}^w(E \cap B_r(a))}{w(B_r(a))}\right)^{p'-1}\frac{dr}{r} \ge \int_0^{t_0}\left(\frac{r^{\alpha p}R_{\alpha,p;1}^w(E \cap B_r(a))}{w(B_r(a))}\right)^{p'-1}\frac{dr}{r}$$

$$\ge \frac{Cl}{2}\int_0^{t_0}\left(\frac{r^{\alpha p}R_{\alpha,p;1}^w(B_r(a))}{w(B_r(a))}\right)^{p'-1}\frac{dr}{r}$$

and the last integral, according to Lemma 3.6.10, diverges. Hence, we conclude that (3.6.12) holds. $\square$

Proof of Lemma 3.6.10. Let ψ and Ψ be as in the proof of the previous lemma. Since $R_{\alpha,p;1}^w(B_t(a))^{p'-1}$, according to Theorem 3.3.10, is comparable to $\Psi(t)^{-1}$ for $0 < t \le 1$, it suffices to show that

$$\int_0^1 \frac{\psi(t)}{\Psi(t)}\, dt = \infty.$$

But this integral is readily computed, viz.

$$\int_0^1 \frac{\psi(t)}{\Psi(t)}\, dt = \log \Psi(0) - \log \Psi(1) = \infty,$$

since

$$\Psi(0) = \int_0^2 \left(\frac{t^{\alpha p}}{w(B_t(a))} \right)^{p'-1} \frac{dt}{t} = \infty$$

by assumption, and

$$0 < \Psi(1) = \int_1^2 \left(\frac{t^{\alpha p}}{w(B_t(a))} \right)^{p'-1} \frac{dt}{t} < \infty. \quad \square$$

The core of the proof of Theorem 3.6.3 is contained in the lemma below, cf. Adams–Meyers [8, pp. 192–193] and Hedberg–Wolff [58, p. 182].

Lemma 3.6.11. *Let $0 < \alpha < N$ and $1 < p < \infty$, and let w be an A_p weight. Let $E \subset \mathbf{R}^N$, and suppose that $a \in \overline{E} \cap e^w_{\alpha,p;1}(E)$. Let $\varepsilon > 0$, and suppose that $a \in B$ for some open ball B. Then*

$$\mathcal{V}^{\mu^{E \cap B}}(a) < \varepsilon, \tag{3.6.14}$$

if B is small enough. Here, $\mu^{E \cap B}$ denotes the capacitary measure for $E \cap B$.

Proof. We first assume that E is a Suslin set. For notational convenience, the measure $\mu^{E \cap B}$ will be denoted by just μ. By (3.6.7), we have

$$\mathcal{V}^{\mu}(a) \le C_2 W_2^{\mu}(a) = C_2 \int_0^2 \left(\frac{t^{\alpha p} \mu(B_t(a))}{w(B_t(a))} \right)^{p'-1} \frac{dt}{t}$$

$$= C_2 \int_0^\delta \left(\frac{t^{\alpha p} \mu(B_t(a))}{w(B_t(a))} \right)^{p'-1} \frac{dt}{t}$$

$$+ C_2 \int_\delta^2 \left(\frac{t^{\alpha p} \mu(B_t(a))}{w(B_t(a))} \right)^{p'-1} \frac{dt}{t}, \tag{3.6.15}$$

where $0 < \delta \le 1$ is arbitrary. Now choose δ so small that

$$\int_0^\delta \left(\frac{t^{\alpha p} R^w_{\alpha,p;1}(E \cap B_t(a))}{w(B_t(a))} \right)^{p'-1} \frac{dt}{t} < \varepsilon',$$

where the value of ε' will be specified shortly. It follows from Lemma 3.6.8 that

$$\mu(B_t(a)) \le R^w_{\alpha,p;1}(E \cap B \cap B_t(a)) \le R^w_{\alpha,p;1}(E \cap B_t(a))$$

for every $t > 0$. This implies that the integral from 0 to δ in the last member of (3.6.15) is less than ε'. Now suppose that $B \subset B_\delta(a)$. Another application of Lemma 3.6.8 shows that

$$\mu(B_t(a)) \le R^w_{\alpha,p;1}(E \cap B \cap B_t(a)) \le R^w_{\alpha,p;1}(E \cap B_\delta(a))$$

for every $t > 0$. If we use this observation together with Theorem 3.3.10, we obtain

$$\int_\delta^2 \left(\frac{t^{\alpha p}\mu(B_t(a))}{w(B_t(a))}\right)^{p'-1} \frac{dt}{t} \le R^w_{\alpha,p;1}(E \cap B_\delta(a))^{p'-1} \int_\delta^2 \left(\frac{t^{\alpha p}}{w(B_t(a))}\right)^{p'-1} \frac{dt}{t}$$

$$\le C\left(\frac{R^w_{\alpha,p;1}(E \cap B_\delta(a))}{R^w_{\alpha,p;1}(B_\delta(a))}\right)^{p'-1}.$$

According to Lemma 3.6.9, we may now choose δ so small that the last quotient is less than $C_2\varepsilon'$. We thus have

$$\mathcal{V}^\mu(a) < 2C_2\varepsilon',$$

so the assertion follows by taking $\varepsilon' = \varepsilon/2C_2$.

Now suppose that E is a general set. A consequence of the doubling property of w is the fact that the *Wiener integral* in the definition of $R^w_{\alpha,p;1}$-thinness converges if and only if

$$\sum_{n=0}^\infty \left(\frac{2^{-\alpha pn}R^w_{\alpha,p;1}(E \cap B_{2^{-n}}(a))}{w(B_{2^{-n}}(a))}\right)^{p'-1} < \infty.$$

We choose, for $n = 1, 2, \ldots$, G_δ sets $F_n \supset E \cap B_{2^{-n}}(a)$ such that

$$R^w_{\alpha,p;1}(F_n) = R^w_{\alpha,p;1}(E \cap B_{2^{-n}}(a)),$$

and then define $F = \bigcap_{n=0}^\infty F_n \setminus B_{2^{-n}}(a)$. The set F is G_δ and

$$R^w_{\alpha,p;1}(F \cap B_{2^{-n}}(a)) = R^w_{\alpha,p;1}(E \cap B_{2^{-n}}(a))$$

for every n, so F is $R^w_{\alpha,p;1}$-thin at a. Let σ be the capacitary measure for $F \cap B$. It is not difficult to see that $\mathcal{V}^\mu(a) = \mathcal{V}^\sigma(a)$. The lemma thus follows from the first part of the proof. □

We are now in a position to prove the Kellogg property of our definition of thinness.

Proof of Theorem 3.6.3. Let $\{B_n\}_{n=1}^\infty$ be an enumeration of the open balls in $\mathbf{R}^N$ with rational centers and radii. Let $\mu_n = \mu^{E \cap B_n}$, and set

$$E_n = \{x \in E \cap B_n \,;\, \mathcal{V}^{\mu_n}(x) < 1\}$$

for $n = 1, 2, \ldots$. It then follows from Proposition 3.6.4 that $R^w_{\alpha,p;1}(E_n) = 0$. By the preceding lemma, we also have $E \cap e^w_{\alpha,p;1}(E) \subset \bigcup_{n=1}^\infty E_n$, which proves the theorem. □

3.6.4. A concept of thinness based on a condensor capacity

A slightly different way of measuring the "thinness" of a set at a point occurs when Poincaré inequalities are employed to prove various results on spectral synthesis; see Section 4.4. It is known that a set E is $R_{\alpha,p;1}$-thin at a point a if and only if

$$\int_0^1 \left(\frac{R_{\alpha,p;t}(E \cap B_t(a))}{t^{N-\alpha p}} \right)^{p'-1} \frac{dt}{t} < \infty;$$

see Adams–Hedberg [7, pp. 233-234]. We intend here to investigate to what extent the corresponding equivalence remains true in the weighted case, and begin by considering the "only if"-part.

Proposition 3.6.12. *Let* $0 < \alpha < N$ *and* $1 < p < \infty$, *and let* w *be a* A_p *weight. Suppose that* $a \in e^w_{\alpha,p;1}(E)$ *for some Suslin set* $E \subset \mathbf{R}^N$. *Then*

$$\int_0^1 \left(\frac{t^{\alpha p} R^w_{\alpha,p;t}(E \cap B_t(a))}{w(B_t(a))} \right)^{p'-1} \frac{dt}{t} < \infty. \tag{3.6.16}$$

Proof. It suffices to show that there exists a constant C such that if t is small enough, then

$$R^w_{\alpha,p;t}(E \cap B_t(a)) \leq C R^w_{\alpha,p;1}(E \cap B_t(a)).$$

We may obviously assume that the left-hand side of this inequality is positive. Let $0 < t \leq \frac{1}{2}$, and let $\mu \in \mathcal{M}^+(E \cap B_t(a))$ be a probability measure, satisfying

$$\|\mathcal{I}_{\alpha,3t}\mu\|_{L^{p'}_{w'}(\mathbf{R}^N)} \leq 2 R^w_{\alpha,p;3t}(E \cap B_t(a))^{-1/p};$$

such a measure exists according to the identity (3.2.3). Another application of this identity shows that

$$R^w_{\alpha,p;1}(E \cap B_t(a))^{-1/p} \leq \|\mathcal{I}_{\alpha,1}\mu\|_{L^{p'}_{w'}(\mathbf{R}^N)}.$$

We also have

$$\|\mathcal{I}_{\alpha,1}\mu\|_{L^{p'}_{w'}(\mathbf{R}^N)} \leq \left(\int_{|x-a|<2t} (\mathcal{I}_{\alpha,1}\mu)^{p'} w' \, dx \right)^{1/p'}$$
$$+ \left(\int_{|x-a|\geq 2t} (\mathcal{I}_{\alpha,1}\mu)^{p'} w' \, dx \right)^{1/p'},$$

and by Minkowski's inequality,

$$\left(\int_{|x-a|\geq 2t} (\mathcal{I}_{\alpha,1}\mu)^{p'} w' \, dx \right)^{1/p'}$$
$$\leq \int_{|y-a|<t} \left(\int_{|x-y|<1, |x-a|\geq 2t} \frac{w'(x)}{|x-y|^{(N-\alpha)p'}} \, dx \right)^{1/p'} d\mu(y).$$

We will now estimate the last integral. Suppose that $|y - a| < t$. Notice that $|x - y| \geq \frac{1}{2}|x - a|$, if $|x - a| \geq 2t$, and that $|x - a| < \frac{3}{2}$, if $|x - y| < 1$. Using these observations and (2.4.8), we find

$$\int_{|y-a|<t} \left(\int_{|x-y|<1, |x-a|\geq 2t} \frac{w'(x)}{|x-y|^{(N-\alpha)p'}} \, dx \right)^{1/p'} d\mu(y)$$

$$\leq 2^{N-\alpha} \mu(E \cap B_t(a)) \left(\int_{2t \leq |x-a|<2} \frac{w'(x)}{|x-a|^{(N-\alpha)p'}} \, dx \right)^{1/p'}$$

$$= 2^{N-\alpha} \left(\int_{2t}^{2} \frac{w'(B_s(a))}{s^{(N-\alpha)p'}} \frac{ds}{s} + \frac{w'(B_2(a))}{2^{(N-\alpha)p'}} - \frac{w'(B_{2t}(a))}{(2t)^{(N-\alpha)p'}} \right)^{1/p'}$$

$$\leq C \left(\int_{2t}^{2} \frac{w'(B_s(a))}{s^{(N-\alpha)p'}} \frac{ds}{s} \right)^{1/p'} .$$

To obtain the last inequality, we have used the fact that

$$\int_{2t}^{2} \frac{w'(B_s(a))}{s^{(N-\alpha)p'}} \frac{ds}{s} \geq \int_{1}^{2} \frac{w'(B_s(a))}{s^{(N-\alpha)p'}} \frac{ds}{s} \geq C \frac{w'(B_1(a))}{2^{(N-\alpha)p'}} \geq C \frac{w'(B_2(a))}{2^{(N-\alpha)p'}},$$

which is a consequence of the doubling property of w. It now follows from Theorem 3.3.10, Lemma 3.6.9, and the assumption that

$$\left(\int_{2t}^{2} \frac{w'(B_s(a))}{s^{(N-\alpha)p'}} \frac{ds}{s} \right)^{1/p'} \leq C \left(\int_{t}^{1} \frac{w'(B_s(a))}{s^{(N-\alpha)p'}} \frac{ds}{s} \right)^{1/p'} \leq C R_{\alpha,p;1}^{w}(B_t(a))^{-1/p}$$

$$= \left(\frac{R_{\alpha,p;1}^{w}(E \cap B_t(a))}{R_{\alpha,p;1}^{w}(B_t(a))} \right)^{1/p} R_{\alpha,p;1}^{w}(E \cap B_t(a))^{-1/p}$$

$$\leq \frac{1}{2} R_{\alpha,p;1}^{w}(E \cap B_t(a))^{-1/p}$$

for t sufficiently close to 0. We have thus shown that

$$R_{\alpha,p;1}^{w}(E \cap B_t(a))^{-1/p} \leq 2 \left(\int_{|x-a|<2t} (\mathcal{I}_{\alpha,1}\mu)^{p'} w' \, dx \right)^{1/p'}, \tag{3.6.17}$$

if t is small enough. If $|x - a| < 2t$ and $|y - a| < t$, then $|x - y| < 3t$. Hence

$$\int_{|x-a|<2t} (\mathcal{I}_{\alpha,1}\mu)^{p'} w' \, dx \leq \int_{\mathbf{R}^N} (\mathcal{I}_{\alpha,3t}\mu)^{p'} w' \, dx$$

$$\leq 2^{p'} R_{\alpha,p;3t}^{w}(E \cap B_t(a))^{-p'/p}.$$

Together with the inequality (3.6.17) this shows that

$$R_{\alpha,p;3t}^{w}(E \cap B_t(a)) \leq C R_{\alpha,p;1}^{w}(E \cap B_t(a)).$$

But by Proposition 3.3.8, $R^w_{\alpha,p;t}(E \cap B_t(a)) \leq CR^w_{\alpha,p;3t}(E \cap B_t(a))$. $\square$

How about the converse statement, i.e., if the integral in (3.6.16) converges, is it true that E is $R^w_{\alpha,p;1}$-thin at a? Well, since $R^w_{\alpha,p;1}(E \cap B_t(a)) \leq R^w_{\alpha,p;t}(E \cap B_t(a))$ for $0 < t \leq 1$,

$$\int_0^1 \left(\frac{t^{\alpha p} R^w_{\alpha,p;1}(E \cap B_t(a))}{w(B_t(a))} \right)^{p'-1} \frac{dt}{t} < \infty,$$

but, as the example below shows, we may not conclude that

$$\int_0^1 \left(\frac{t^{\alpha p}}{w(B_t(a))} \right)^{p'-1} \frac{dt}{t} = \infty.$$

However, if we assume that a belongs to E, then the converse holds. This is the content of Proposition 3.6.14. The proof resembles very much that of Lemma 3.6.9.

Example 3.6.13. Let w be an A_p weight such that $w(x)$ is continuous for $x \neq a$ and

$$\int_0^1 \left(\frac{t^{\alpha p}}{w(B_t(a))} \right)^{p'-1} \frac{dt}{t} < \infty; \tag{3.6.18}$$

we can for instance take $w(x) = |x|^{-\eta}$ and $a = 0$, where the exponent η is chosen so that $0 < \eta < N$ and $\alpha p + \eta > N$. Now let $\{x_i\}_{i=1}^\infty$ be any sequence of points converging to a such that $x_i \neq a$ for every i, and define $E = \bigcup_{i=1}^\infty \{x_i\}$. It is not difficult to show that if $0 < t < 1$, then the capacity $R^w_{\alpha,p;t}(\{x_i\})$ is comparable to

$$\left(\int_0^t \left(\frac{s^{\alpha p}}{w(B_s(x_i))} \right)^{p'-1} \frac{ds}{s} \right)^{1-p};$$

see D. R. Adams [5, pp. 83–84]. But this integral diverges because

$$\fint_{B_s(x_i)} w \, dy \to w(x_i), \quad \text{as } s \to 0,$$

and $w(x_i) < \infty$. Thus, $R^w_{\alpha,p;t}(\{x_i\}) = 0$, which implies that $R^w_{\alpha,p;t}(E) = 0$. This shows that the integral in (3.6.16) vanishes. But E is $R_{\alpha,p;1}$-thick at a because of the assumption (3.6.18).

Proposition 3.6.14. *Let $0 < \alpha < N$ and $1 < p < \infty$, and let w be an A_p weight. Let E be a subset of $\mathbf{R}^N$, and let a be a point belonging to E such that*

$$\int_0^1 \left(\frac{t^{\alpha p}}{w(B_t(a))} \right)^{p'-1} \frac{dt}{t} < \infty.$$

Then

$$\int_0^1 \left(\frac{t^{\alpha p} R^w_{\alpha,p;t}(E \cap B_t(a))}{w(B_t(a))} \right)^{p'-1} \frac{dt}{t} = \infty. \tag{3.6.19}$$

Proof. If we define the function ψ by

$$\psi(t) = \left(\frac{t^{\alpha p}}{w(B_t(a))}\right)^{p'-1}\frac{1}{t}, \quad 0 < t \le 1,$$

as in Lemma 3.6.9, then ψ is continuous on $(0, 1]$ and $\int_0^1 \psi(t)\,dt < \infty$. Moreover, if $0 < \rho \le 1$, then

$$\int_\rho^1 \frac{\psi(t)}{\int_0^t \psi(r)\,dr}\,dt = \log\int_0^1 \psi(t)\,dt - \log\int_0^\rho \psi(t)\,dt.$$

As $\rho \to 0$, the right-hand side of this identity tends to ∞. But in view of the fact that $R_{\alpha,p;t}^w(\{a\})^{p'-1}$ by Theorem 3.3.10 is comparable to $\left(\int_0^t \psi(r)\,dr\right)^{-1}$, this shows that

$$\int_0^1 \left(\frac{t^{\alpha p}R_{\alpha,p;t}^w(\{a\})}{w(B_t(a))}\right)^{p'-1}\frac{dt}{t} = \infty,$$

and the proposition follows since $a \in E$. $\square$

Remark 3.6.15. "The Kellogg lemma" (Theorem 3.6.3) and the proposition just proved together imply that the set of points $a \in E$ for which (3.6.19) does not hold, has $R_{\alpha,p;1}^w$-capacity 0.

3.7. Thinness: The case $p = 1$

In this section, we propose a definition of thinness for the case $p = 1$.[†] By using the same ideas as in the non-weighted case, it is possible to prove the following theorem; see Vodop'yanov [104].[‡]

Theorem 3.7.1. Let $0 < \alpha < N$ and $1 < p < \infty$, and let $w \in A_p$. Then a set $E \subset \mathbf{R}^N$ is $R_{\alpha,p;1}^w$-thin at a point $a \in \overline{E}$ if and only if $R_{\alpha,p;1}^w(\{a\}) = 0$ and there exists a measure $\mu \in \mathcal{M}^+(\mathbf{R}^N)$ such that

$$\liminf_{x \to a,\ x \in E\setminus\{a\}} W_1^\mu(x) > W_1^\mu(a).$$

Here, W_1^μ is the potential defined in (3.6.6):

$$W_1^\mu(x) = \int_0^1 \left(\frac{t^{\alpha p}\mu(B_t(x))}{w(B_t(x))}\right)^{p'-1}\frac{dt}{t}, \quad x \in \mathbf{R}^N.$$

[†]The reader should notice that the condition (3.6.2) degenerates, if we first raise the left hand side to $p - 1$ and then let $p \to 1$.

[‡]In [104], this is proved for the homogeneous Riesz capacity $R_{\alpha,p}^w$.

Notice that $W_1^\mu(x)^{p-1} \to M_{w,1}^{N-\alpha}\mu(x)$, as $p \to 1$, where

$$M_{w,1}^{N-\alpha}\mu(x) = \sup_{0 < r \le 1} \frac{\mu(B_r(x))}{h_w^{N-\alpha}(B_r(x))}$$

is the fractional maximal function that was introduced in Section 3.4 in connection with Frostman's lemma. As before, $h_w^{N-\alpha}(B_r(x)) = r^{-\alpha}w(B_r(x))$. We shall also use another maximal function.

Definition 3.7.2. Let $0 < \alpha < N$ and $0 < r \le \infty$, and let w be a weight. For a measure $\mu \in \mathcal{M}^+(\mathbf{R}^N)$, we define the fractional maximal function $\mathcal{M}_{w,r}^{N-\alpha}\mu$ by

$$\mathcal{M}_{w,r}^{N-\alpha}\mu(x) = \sup_{0 < s < r} \sup_{|x-y| < s} \frac{\mu(B_s(y))}{h_w^{N-\alpha}(B_s(y))}$$

for $x \in \mathbf{R}^N$.

It is easy to see that

$$\mathcal{M}_{w,r}^{N-\alpha}\mu(x) \le C M_{w,2r}^{N-\alpha}\mu(x), \tag{3.7.1}$$

if w is doubling and that both maximal functions are lower semicontinuous as functions of x.

Definition 3.7.3. Let $0 < \alpha < N$, and let w be a weight. We say that a set $E \subset \mathbf{R}^N$ is $\mathcal{H}_{w,1}^{N-\alpha}$-*thin* at a point $a \in \mathbf{R}^N$ if $\mathcal{H}_{w,1}^{N-\alpha}(\{a\}) = 0$ and there exists a measure $\mu \in \mathcal{M}^+(\mathbf{R}^N)$ and a positive constant η such that

$$\liminf_{x \to a, \ x \in E \setminus \{a\}} \mathcal{M}_{w,r}^{N-\alpha}\mu(x) \ge \mathcal{M}_{w,r}^{N-\alpha}\mu(a) + \eta \tag{3.7.2}$$

for every r such that $0 < r < 1$, and, furthermore,

$$\lim_{r \to 0} \mathcal{M}_{w,r}^{N-\alpha}\mu(a) = 0. \tag{3.7.3}$$

The set of points where E is $\mathcal{H}_{w,1}^{N-\alpha}$-thin is denoted $e_{\alpha,1;1}^w(E)$.

Remark 3.7.4. Notice that the assumption $\mathcal{H}_{w,1}^{N-\alpha}(\{a\}) = 0$ corresponds to the condition (3.6.1) in Definition 3.6.1. The condition that (3.7.2) should hold for every r such that $0 < r < 1$ guarantees that the discontinuity of the maximal functions comes from properties of E arbitrarily close to a. The assumption (3.7.3) is imposed for technical reasons. The author does not know whether Theorem 3.7.5 below is true if (3.7.3) is removed from the definition.

The following theorem shows that this definition of thinness has the Kellogg property.

Theorem 3.7.5. *Let $0 < \alpha < N$, and let $w \in A_1$. If $E \subset \mathbf{R}^N$, then*

$$\mathcal{H}_{w,1}^{N-\alpha}(E \cap e_{\alpha,1;1}^w(E)) = 0.$$

For the proof of this theorem, we need two lemmas that both concern $\mathcal{H}_{w,1}^{N-\alpha}$-capacitary densities and may be of independent interest.

Definition 3.7.6. Let $0 < \alpha < N$, and let w be a weight. If $E \subset \mathbf{R}^N$, we define the *upper $\mathcal{H}_{w,1}^{N-\alpha}$-density* of E at a point $x \in \mathbf{R}^N$, $\overline{D}_{w,1}^{N-\alpha}(x,E)$, by

$$\overline{D}_{w,1}^{N-\alpha}(x,E) = \lim_{r \to 0} \sup_{0 < s < r} \sup_{|x-y| < s} \frac{\mathcal{H}_{w,1}^{N-\alpha}(E \cap B_s(y))}{h_w^{N-\alpha}(B_s(y))}.$$

Note that $\overline{D}_{w,1}^{N-\alpha}(x,E) \le 1$ for every $x \in \mathbf{R}^N$.

Lemma 3.7.7. *Let $0 < \alpha < N$, and let $w \in A_1$. If $E \subset \mathbf{R}^N$, then*

$$\mathcal{H}_{w,1}^{N-\alpha}(\{x \in E \; ; \; \overline{D}_{w,1}^{N-\alpha}(x,E) < 1\}) = 0.$$

The corresponding result for non-weighted Hausdorff measures is due to S. Kametani [64, p. 618]. Kametani's result was later extended to non-weighted Hausdorff capacities by C. Fernström [40, pp. 27–28] and to weighted Hausdorff capacities by E. Nieminen [91, p. 22]. Nieminen only assumes that the weight w is doubling but instead that the measure function $h_w^{N-\alpha}(B_r(x))$ essentially is increasing as a function of r. This assumption is not satisfied by all A_1 weights.

Remark 3.7.8.

1. Since

$$\mathcal{H}_{w,1}^{N-\alpha}(B_s(y)) \le \frac{w(B_s(y))}{s^\alpha}$$

   for every ball $B_s(y)$, the lemma implies that the set of points $x \in E$, where the density

$$\lim_{r \to 0} \sup_{0 < s < r} \sup_{|x-y| < s} \frac{\mathcal{H}_{w,1}^{N-\alpha}(E \cap B_s(y))}{\mathcal{H}_{w,1}^{N-\alpha}(B_s(y))} < 1,$$

   also has capacity 0.

2. Notice that it follows from the lemma that if $a \in E$ and $\mathcal{H}_{w,1}^{N-\alpha}(\{a\}) > 0$, then $\overline{D}_{w,1}^{N-\alpha}(a,E) = 1$.

The proof is a modification of the proof in Fernström [40].

Proof of Lemma 3.7.7. We define $E_m = \{x \in E \; ; \; \overline{D}(x,E) < 1 - 1/m\}$ for $m = 2, 3, \ldots$, and then let E_{mn}, $n = 1, 2, \ldots$, be the set of $x \in E_m$, satisfying

$$\mathcal{H}_{w,1}^{N-\alpha}(E \cap B_s(y)) < \left(1 - \frac{1}{m}\right) h_w^{N-\alpha}(B_s(y)) \qquad (3.7.4)$$

for every ball $B_s(y)$ with $x \in B_s(y)$ and $s < 1/n$. We then see that the inclusion $\{x \in E \, ; \, \overline{D}(x, E) < 1\} \subset \bigcup_{m,n} E_{mn}$ holds, and it is enough to show that $\mathcal{H}^{N-\alpha}_{w,1}(E_{mn}) = 0$ for every m and n under consideration. Suppose that $\mathcal{H}^{N-\alpha}_{w,1}(E_{mn}) > 0$ for some m and n. Let $x \in \mathbf{R}^N$ satisfy $0 < w(x) < \infty$ and

$$\fint_{B_r(x)} w \, dy \to w(x), \quad \text{as } r \to 0.$$

Let $0 < r < 1/n$. If $B_r(x) \cap B_{r'}(x')$ is non-empty for some ball $B_{r'}(x')$, where $1/n \le r' \le 1$, then $B_{1/n}(x) \subset B_{3r'}(x')$. By the doubling property of w and Lemma 3.4.4, we then have

$$\frac{w(B_{r'}(x'))}{r'^\alpha} \ge C \frac{w(B_{3r'}(x'))}{(3r')^\alpha} \ge A \inf_{1/n \le t \le 1} \frac{w(B_t(x))}{t^\alpha}.$$

Since

$$r^{N-\alpha} \fint_{B_r(x)} w \, dy \to 0, \quad \text{as } r \to 0,$$

it is possible to choose $r = r(x)$ such that $0 < r < 1/n$ and

$$\left(1 + \frac{1}{m}\right) \frac{w(B_r(x))}{r^\alpha} = C\left(1 + \frac{1}{m}\right) r^{N-\alpha} \fint_{B_r(x)} w \, dy$$

$$< A \inf_{1/n \le t \le 1} \frac{w(B_t(x))}{t^\alpha}. \tag{3.7.5}$$

We thus have

$$\left(1 + \frac{1}{m}\right) h^{N-\alpha}_w(B_r(x)) < h^{N-\alpha}_w(B_{r'}(x')) \tag{3.7.6}$$

for every ball $B_{r'}(x')$ with radius $1/n \le r' \le 1$, that intersects $B_r(x)$. Let $\{B_j\}$ be a countable covering of $\mathbf{R}^N$ with balls B_j, satisfying (3.7.5). It then follows that $\mathcal{H}^{N-\alpha}_{w,1}(E_{mn} \cap B_k) > 0$ for some ball $B_k \in \{B_j\}$. Suppose that $\{B'_j\}$ is a covering of $E_{mn} \cap B_k$ so that $B'_j \cap (E_{mn} \cap B_k) \ne \emptyset$ for every j, each B'_j has radius ≤ 1, and

$$\sum_j h^{N-\alpha}_w(B'_j) < \left(1 + \frac{1}{m}\right) \mathcal{H}^{N-\alpha}_{w,1}(E_{mn} \cap B_k).$$

Then each ball B'_j has to have radius less than $1/n$, for if the radius of some B'_i is $\ge 1/n$, then by (3.7.6),

$$\left(1 + \frac{1}{m}\right) \mathcal{H}^{N-\alpha}_{w,1}(E_{mn} \cap B_k) \le \left(1 + \frac{1}{m}\right) \mathcal{H}^{N-\alpha}_{w,1}(B_k) \le \left(1 + \frac{1}{m}\right) h^{N-\alpha}_w(B_k)$$

$$< h^{N-\alpha}_w(B'_i) \le \sum_j h^{N-\alpha}_w(B'_j)$$

$$< \left(1 + \frac{1}{m}\right) \mathcal{H}^{N-\alpha}_{w,1}(E_{mn} \cap B_k),$$

which is impossible. By first using the subadditivity of $\mathcal{H}_{w,1}^{N-\alpha}$ and then the property (3.7.4), we obtain

$$\mathcal{H}_{w,1}^{N-\alpha}(E_{mn} \cap B_k) \le \sum_j \mathcal{H}_{w,1}^{N-\alpha}(E_{mn} \cap B_k \cap B'_j)$$

$$\le \sum_j \mathcal{H}_{w,1}^{N-\alpha}(E \cap B'_j)$$

$$< \left(1 - \frac{1}{m}\right) \sum_j h_w^{N-\alpha}(B'_j)$$

$$< \left(1 - \frac{1}{m^2}\right) \mathcal{H}_{w,1}^{N-\alpha}(E_{mn} \cap B_k),$$

again a contradiction. We conclude that $\mathcal{H}_{w,1}^{N-\alpha}(E_{mn}) = 0$ for every m and n, and this proves the theorem. $\square$

The next lemma is a counterpart to Lemma 3.6.9 for $p = 1$.

Lemma 3.7.9. *Let $0 < \alpha < N$, and let $w \in A_1$. If a set $E \subset \mathbf{R}^N$ is $\mathcal{H}_{w,1}^{N-\alpha}$-thin at a point $a \in \overline{E}$, then $\overline{D}_{w,1}^{N-\alpha}(a, E) = 0$.*

Theorem 3.7.5 follows immediately from Lemma 3.7.7 and Lemma 3.7.9. The proof of Lemma 3.7.9 utilizes a weak capacitary inequality for $\mathcal{H}_{w,1}^{N-\alpha}$.

Lemma 3.7.10. *Let $0 < \alpha < N$, and let $w \in A_1$. Then there exists a constant C, which only depends on α, N, and the A_1 constant of w, such that, for every measure $\mu \in \mathcal{M}^+(\mathbf{R}^N)$,*

$$\mathcal{H}_{w,1}^{N-\alpha}(\{x \,;\, \mathcal{M}_{w,r}^{N-\alpha}\mu(x) > \lambda\}) \le \frac{C}{\lambda}\mu(\mathbf{R}^N)$$

for every r such that $0 < r < \frac{1}{10}$ and every $\lambda > 0$.

Remark 3.7.11. It follows from the lemma that the maximal function $\mathcal{M}_{w,r}^{N-\alpha}\mu$ is finite $\mathcal{H}_{w,1}^{N-\alpha}$-q.e.

The proof is inspired by Adams–Hedberg [7, pp. 173–174].

Proof. In view of (3.7.1), it suffices to prove that

$$\mathcal{H}_{w,1}^{N-\alpha}(\{x \,;\, M_{w,2r}^{N-\alpha}\mu(x) > \lambda\}) \le \frac{C}{\lambda}\mu(\mathbf{R}^N) \tag{3.7.7}$$

for every r, $0 < r < \frac{1}{10}$, and every $\lambda > 0$. For fixed r and λ, let E_λ be the open set $\{x \,;\, M_{w,2r}^{N-\alpha}\mu(x) > \lambda\}$. We first assume that the left-hand side in (3.7.7) is finite. It follows from the capacitability theorem (Theorem 3.4.25) and

Frostman's lemma (Theorem 3.4.27) that there exists a compact set $K \subset E_\lambda$ and a measure $\gamma \in \mathcal{M}^+(K)$ so that

$$\mathcal{H}_{w,10r}^{N-\alpha}(E_\lambda) \leq 2\,\mathcal{H}_{w,10r}^{N-\alpha}(K) \leq C\gamma(K)$$

and $\mathcal{M}_{w,10r}^{N-\alpha}\gamma(x) \leq 1$ for every $x \in \mathbf{R}^N$. If we define

$$M_\gamma\mu(x) = \sup_{0<s<10r} \frac{\mu(B_{s/5}(x))}{\gamma(B_s(x))}$$

for $x \in \mathbf{R}^N$, it follows that

$$\begin{aligned}
M_{w,2r}^{N-\alpha}\mu(x) &= \sup_{0<s<10r} \frac{\mu(B_{s/5}(x))}{\gamma(B_s(x))}\,\frac{\gamma(B_s(x))}{h_w^{N-\alpha}(B_{s/5}(x))} \\
&\leq CM_\gamma\mu(x)\,M_{w,10r}^{N-\alpha}\gamma(x) \\
&\leq CM_\gamma\mu(x),
\end{aligned}$$

so that $\operatorname{supp}\gamma \subset E_\lambda \subset \{x \ ; \ M_\gamma\mu(x) > \lambda/C\}$. Now cover $\operatorname{supp}\gamma$ with balls $\{B_{r_i}(x_i)\}$ such that the balls $\{B_{r_i/5}(x_i)\}$ are disjoint and

$$\gamma(B_{r_i}(x_i)) < \frac{C}{\lambda}\mu(B_{r_i/5}(x_i))$$

for every i (cf. Stein [99, p. 9]). We now obtain the inequality (3.7.7) as follows:

$$\begin{aligned}
\mathcal{H}_{w,1}^{N-\alpha}(E_\lambda) &\leq \mathcal{H}_{w,10r}^{N-\alpha}(E_\lambda) \leq C\gamma(K) \leq C\sum_i \gamma(B_{r_i}(x_i)) \\
&< \frac{C}{\lambda}\sum_i \mu(B_{r_i/5}(x_i)) \leq \frac{C}{\lambda}\mu(\mathbf{R}^N).
\end{aligned}$$

We finally have to consider the case when the left-hand side in (3.7.7) is infinite. For $m = 1, 2, \ldots$, there are compact subsets K_m of E_λ and measures $\gamma_m \in \mathcal{M}^+(K_m)$ such that $\gamma_m(K_m) \geq m$ and $\mathcal{M}_{w,10r}^{N-\alpha}\gamma_m(x) \leq 1$ for every point $x \in \mathbf{R}^N$. Just as before, we get

$$m \leq \gamma_m(K_m) < \frac{C}{\lambda}\mu(\mathbf{R}^N).$$

Since m can be arbitrarily large, this implies that $\mu(\mathbf{R}^N) = \infty$, so (3.7.7) holds in this case as well. $\square$

Proof of Lemma 3.7.9. Let $\varepsilon > 0$ be an arbitrary constant, and choose $0 < r < \frac{1}{10}$ so small that $\mathcal{M}_{w,4r}^{N-\alpha}\mu(a) < \frac{1}{2}\eta\varepsilon$. Let the number $\delta > 0$ satisfy

$$\inf_{|x-a|<\delta,\ x\in E\backslash\{a\}} \mathcal{M}_{w,r}^{N-\alpha}\mu(x) \geq \frac{1}{2}\left(\mathcal{M}_{w,r}^{N-\alpha}\mu(a) + \eta\right),$$

where η is the number in (3.7.2), and let γ be the measure $2\mu/\eta$. We then find that $\mathcal{M}_{w,6r}^{N-\alpha}\gamma(a) < \varepsilon$ and

$$\inf_{|x-a|<\delta,\; x\in E\setminus\{a\}} \mathcal{M}_{w,r}^{N-\alpha}\gamma(x) \geq 1. \tag{3.7.8}$$

If $t > 0$ satisfies $t < r$ and $t < \delta$, we define $\gamma_t = \gamma|_{B_t(a)}$ and $\gamma_t' = \gamma - \gamma_t$. Suppose that $x \in E$ and $|x - a| < \frac{1}{2}t$. If y is such that $|x - y| < s \leq \frac{1}{4}t$, then $B_s(y) \subset B_t(a)$, which implies that $\gamma_t'(B_s(y)) = 0$. It follows that

$$\mathcal{M}_{w,r}^{N-\alpha}\gamma_t'(x) = \sup_{t/4<s<r} \sup_{|x-y|<s} \frac{\gamma_t'(B_s(y))}{h_w^{N-\alpha}(B_s(y))}.$$

Moreover, if $|x - y| < s$, where $s > \frac{1}{4}t$, then $B_s(y) \subset B_{4s}(a)$, and hence

$$\mathcal{M}_{w,r}^{N-\alpha}\gamma_t'(x) \leq \sup_{t/4<s<r} \sup_{|x-y|<s} \frac{\gamma_t'(B_{4s}(a))}{h_w^{N-\alpha}(B_s(y))} \leq C \sup_{t<s<4r} \frac{\gamma_t'(B_s(a))}{h_w^{N-\alpha}(B_s(a))}$$

$$\leq C\mathcal{M}_{w,4r}^{N-\alpha}\gamma(a) < C\varepsilon. \tag{3.7.9}$$

We also have

$$\mathcal{M}_{w,r}^{N-\alpha}\gamma(x) \leq \mathcal{M}_{w,r}^{N-\alpha}\gamma_t(x) + \mathcal{M}_{w,r}^{N-\alpha}\gamma_t'(x).$$

It thus follows from the inequalities (3.7.8) and (3.7.9) that, for every $x \in E\setminus\{a\}$ satisfying $|x - a| < \frac{1}{2}t$,

$$\mathcal{M}_{w,r}^{N-\alpha}\gamma_t(x) \geq 1 - C\varepsilon > \tfrac{1}{2},$$

if ε is so small that $C\varepsilon < \frac{1}{2}$. Thus, we see that $E \setminus \{a\} \cap B_{t/2}(a)$ is a subset of the set $\{x \; ; \; \mathcal{M}_{w,r}^{N-\alpha}\gamma_t(x) > \frac{1}{2}\}$. Using Lemma 3.7.10 and the assumption that $\mathcal{H}_{w,1}^{N-\alpha}(\{a\}) = 0$, we get

$$\mathcal{H}_{w,1}^{N-\alpha}(E \cap B_{t/2}(a)) \leq C\gamma_t(\mathbf{R}^N) = C\gamma(B_t(a)),$$

which implies that

$$\frac{\mathcal{H}_{w,1}^{N-\alpha}(E \cap B_{t/2}(a))}{h_w^{N-\alpha}(B_{t/2}(a))} \leq C\frac{\gamma(B_t(a))}{h_w^{N-\alpha}(B_t(a))} < C\varepsilon,$$

if $0 < t < \delta$ satisfies $t < r$. Hence

$$\lim_{t\to 0} \frac{\mathcal{H}_{w,1}^{N-\alpha}(E \cap B_t(a))}{h_w^{N-\alpha}(B_t(a))} = 0.$$

But if $|x - a| < t$, then $B_t(x) \subset B_{2t}(a)$, so

$$\frac{\mathcal{H}_{w,1}^{N-\alpha}(E \cap B_t(x))}{h_w^{N-\alpha}(B_t(x))} \leq C\frac{\mathcal{H}_{w,1}^{N-\alpha}(E \cap B_{2t}(a))}{h_w^{N-\alpha}(B_{2t}(a))},$$

and, consequently, $\overline{D}_{w,1}^{N-\alpha}(a, E) = 0$. $\quad\square$

Chapter 4

Applications of potential theory to Sobolev spaces

The nonlinear potential theory that was developed in Chapter 3 will be applied to the study of weighted Sobolev spaces in this chapter.

In Section 4.1, we show that every function in $W_w^{m,p}(\mathbf{R}^N)$, where w is an A_p weight, has a Borel measurable representative that is quasicontinuous with respect to the Bessel capacity $B_{m,p}^w$, if $1 < p < \infty$, and the Hausdorff capacity $\mathcal{H}_{w,1}^{N-m}$, if $p = 1$. It follows that functions in $W_w^{m,p}(\Omega)$ are quasicontinuous, if Ω is an extension domain. We also prove uniqueness theorems for quasicontinuous representatives of functions in $W_w^{m,p}(\Omega)$, where the domain Ω satisfies a metric density condition. These theorems generalize earlier results by V. P. Havin and V. G. Maz'ya and by A. Carlsson.

In Section 4.2, sufficient conditions for a measure to belong to the dual of $W_w^{m,p}(\Omega)$ are given. The corresponding results in the non-weighted case are due to A. Carlsson.

A key tool in L. I. Hedberg's proof of the theorem on spectral synthesis mentioned in the preface are so called Poincaré inequalities. In Section 4.3, we obtain weighted Poincaré inequalities using methods developed by N. G. Meyers.

Finally, in Section 4.4, two results on spectral synthesis are established. The first of these concerns the case $1 < p < \infty$ and "uniformly thick" domains. The second result is for $p = 1$ and domains that satisfy a capacitary cone condition. The methods used in this section are due to Hedberg.

4.1. Quasicontinuity

An important and very useful property of the Sobolev space $W^{m,p}(\mathbf{R}^N)$ is the fact that every element of this space, viewed as an equivalence class of functions coinciding a.e., has a distinguished representative which is *quasicontinuous* with respect to some suitable capacity. This means, in particular, that the representative is defined more often than almost everywhere with respect to Lebesgue measure, namely, outside a set of capacity 0, where the exceptional sets are measured by Bessel capacity $B_{m,p}$, if $1 < p < \infty$, and Hausdorff capacity $\mathcal{H}_1^{N-m}$, in the case $p = 1$ (or by some equivalent capacity). It is thus meaningful to talk about the values of a Sobolev function on "small" sets with measure 0 but with positive capacity, for instance, on the boundary of some domain; what is meant are just the values of the quasicontinuous representative. The property of being quasicontinuous may more generally be viewed as the counterpart in potential theory to the familiar Luzin property of measurable functions.

4.1.1. The case $1 < p < \infty$

Let us first define quasicontinuity with respect to the weighted Bessel capacity, that was introduced in Section 3.3.

Definition 4.1.1. Let $0 < \alpha < N$ and $1 < p < \infty$. Let w be a weight. A function f, defined $B_{\alpha,p}^w$-q.e. on an open subset O of $\mathbf{R}^N$, is said to be $B_{\alpha,p}^w$-*quasicontinuous*, if, given $\varepsilon > 0$, there is an open set $G \subset O$ such that $B_{\alpha,p}^w(G) < \varepsilon$ and $f|_{O \setminus G}$ is continuous.

We will first show that the Bessel potentials $\mathcal{G}_\alpha f$ are $B_{\alpha,p}^w$-quasicontinuous. The proof is the same as in Meyers [77, p. 280]. Note that this result and Theorem 3.3.2 together imply that functions in $W_w^{m,p}(\mathbf{R}^N)$ are $B_{m,p}^w$-quasicontinuous, if w is an A_p weight.

Theorem 4.1.2. Let $0 < \alpha < N$ and $1 < p < \infty$. Let w be a weight. Suppose that $f \in L_w^p(\mathbf{R}^N)$. Then $\mathcal{G}_\alpha f$ is $B_{\alpha,p}^w$-quasicontinuous.

Proof. According to Proposition 3.2.2, $\mathcal{G}_\alpha f$ is defined and finite $B_{\alpha,p}^w$-q.e. Let $\{f_n\}_{n=1}^\infty$ be a sequence of functions in $C_0^\infty(\mathbf{R}^N)$ such that f_n converges to f in $L_w^p(\mathbf{R}^N)$ (the existence of such a sequence follows from, e.g., Theorem 3.14 in Rudin [95, p. 71]). Proposition 3.2.24 shows that, for a subsequence $\{f_{n_j}\}_{j=1}^\infty$,

$$\mathcal{G}_\alpha f_{n_j}(x) \to \mathcal{G}_\alpha f(x)$$

for $B_{\alpha,p}^w$-q.e. $x \in \mathbf{R}^N$ and uniformly outside an open set of arbitrarily small capacity. The assertion now follows from the fact that $\mathcal{G}_\alpha f_{n_j} \in \mathcal{S}$. $\square$

Corollary 4.1.3. Let m be an integer, $1 \le m < N$, and let $1 < p < \infty$. Let w be an A_p weight. Then every function in $W_w^{m,p}(\mathbf{R}^N)$ has a Borel measurable, $B_{m,p}^w$-quasicontinuous representative.

We next prove a uniqueness theorem for $B_{\alpha,p}^w$-quasicontinuous functions, a consequence of which is the fact that the $B_{m,p}^w$-quasicontinuous representative of a function belonging to $W_w^{m,p}(\mathbf{R}^N)$ is essentially unique.[†] The proof uses a result concerning limits of mean values for Bessel potentials;[‡] see also Adams–Hedberg [7, p. 161].

Proposition 4.1.4. *Let $0 < \alpha < N$ and $1 < p < \infty$, and let w be a weight. Suppose that $f \in L_w^p(\mathbf{R}^N)$. Then*

$$\lim_{r \to 0} \fint_{B_r(x)} \mathcal{G}_\alpha f(y)\, dy = \mathcal{G}_\alpha f(x)$$

for $B_{\alpha,p}^w$-q.e. $x \in \mathbf{R}^N$.

Proof. Let χ be the characteristic function for the unit ball $B_1(0)$, and define for $r > 0$, $\chi_r(x) = |B_r(0)|^{-1}\chi(x/r)$, $x \in \mathbf{R}^N$. Then

$$\fint_{B_r(x)} \mathcal{G}_\alpha f\, dy = \chi_r * (G_\alpha * f)(x) = (\chi_r * G_\alpha) * f(x)$$

$$= \int_{\mathbf{R}^N} \chi_r * G_\alpha(y) f(x - y)\, dy. \tag{4.1.1}$$

As $r \to 0$, $\chi_r * G_\alpha(y) \to G_\alpha(y)$ for every $y \in \mathbf{R}^N$. This implies that, for fixed $x \in \mathbf{R}^N$, $\chi_r * G_\alpha(y) f(x - y) \to G_\alpha(y) f(x - y)$ for a.e. every $y \in \mathbf{R}^N$. It is easy to show that

$$\chi_r * G_\alpha(y) \le C G_\alpha(y)$$

for $0 < r \le 1$ and $y \in \mathbf{R}^N$; see Adams–Hedberg [7, p. 161]. The integrand in the last integral in (4.1.1) is thus dominated by a constant times $G_\alpha(y)|f(x - y)|$, which, by Proposition 3.2.2, is an L^1 function for $B_{\alpha,p}^w$-q.e. $x \in \mathbf{R}^N$. The proposition now follows from Lebesgue's dominated convergence theorem. $\square$

The uniqueness theorem below contains a slight improvement of a result by A. Carlsson [20, p. 5] for the non-weighted case.[§] Carlsson's theorem generalizes earlier work by V. G. Maz'ya and V. P. Havin [53, p. 117], that concerns $R_{\alpha,p}$-quasicontinuous functions coinciding a.e. on $\mathbf{R}^N$. A weighted version of Maz'ya and Havin's uniqueness theorem has been obtained by H. Aikawa [10, p. 344].

Theorem 4.1.5. *Let $0 < \alpha < N$ and $1 < p < \infty$, and let w be a weight. Suppose that O is an open subset of $\mathbf{R}^N$ such that*

$$\limsup_{r \to 0} \frac{|O \cap B_r(x)|}{|B_r(x)|} > 0 \tag{4.1.2}$$

[†]Similar results have been obtained by J. Heinonen, T. Kilpeläinen, and O. Martio [59, p. 95]. See also Kilpeläinen [67, p. 110].

[‡]This result is stated, but not proved in Aikawa [10, p. 344].

[§]Carlsson assumes that there is a constant $a > 0$ such that

$$\liminf_{r \to 0} \frac{|O \cap B_r(x)|}{|B_r(x)|} \ge a$$

for $B_{\alpha,p}$-q.e. $x \in \partial O$.

for $B^w_{\alpha,p}$-q.e. $x \in \partial O$. Let f_1 and f_2 be two $B^w_{\alpha,p}$-quasicontinuous functions on $\mathbf{R}^N$ such that $f_1 = f_2$ a.e. on O. Then $f_1 = f_2$ $B^w_{\alpha,p}$-q.e. on $\bar{O}$.

Proof. The function $f = f_1 - f_2$ is $B^w_{\alpha,p}$-quasicontinuous, so there exist open sets G_n, $n = 1, 2, \ldots$, such that $B^w_{\alpha,p}(G_n) \to 0$, as $n \to \infty$, and $f|_{G^c_n}$ is continuous. Let ψ_n be nonnegative functions in $L^p_w(\mathbf{R}^N)$, satisfying $\|\psi_n\|_{L^p_w(\mathbf{R}^N)} \to 0$ and $\varphi_n = \mathcal{G}_\alpha \psi_n \geq 1$ on G_n. According to Proposition 3.2.24, we can assume that $\varphi_n \to 0$ $B^w_{\alpha,p}$-q.e. by passing to a subsequence.

Let $x \in \bar{O}$ be a point for which $\varphi_n(x) \to 0$ and the density in (4.1.2) is positive. We then obtain

$$\limsup_{r \to 0} \frac{|G_n \cap B_r(x)|}{|B_r(x)|} \leq \limsup_{r \to 0} \fint_{B_r(x)} \chi_{G_n(y)}(x)\varphi_n(y)\,dy$$

$$\leq \lim_{r \to 0} \fint_{B_r(x)} \varphi_n(y)\,dy = \varphi_n(x).$$

Hence, because $\varphi_n(x) \to 0$, we can find n such that

$$\limsup_{r \to 0} \frac{|G_n \cap B_r(x)|}{|B_r(x)|} < \limsup_{r \to 0} \frac{|O \cap B_r(x)|}{|B_r(x)|}.$$

This fact implies that $|O \cap B_r(x) \cap G^c_n| > 0$ for arbitrarily small r, so we can choose a sequence $\{x_j\}^\infty_{n=1}$ of points belonging to $O \cap G^c_n$ such that $x_j \to x$, as $j \to \infty$, and $f(x_j) = 0$ for every j. Since $x \in G^c_n$, it follows that $f(x) = 0$, which is exactly what we wanted to prove. $\quad\square$

4.1.2. The case $p = 1$

We now turn our attention to the case $p = 1$. The definition of quasicontinuity with respect to Hausdorff capacity is completely analogous to the definition in the previous section.

Definition 4.1.6. Let $0 < \alpha < N$. Let w be a weight. A function f, defined $\mathcal{H}^{N-\alpha}_{w,1}$-q.e. on an open subset O of $\mathbf{R}^N$, is said to be $\mathcal{H}^{N-\alpha}_{w,1}$-*quasicontinuous*, if, given $\varepsilon > 0$, there is an open set $G \subset O$ such that $\mathcal{H}^{N-\alpha}_{w,1}(G) < \varepsilon$ and $f|_{O \setminus G}$ is continuous.

We will first show the existence of $\mathcal{H}^{N-m}_{w,1}$-quasicontinuous representatives for functions in $W^{m,1}_w(\mathbf{R}^N)$. Recall that we in Section 3.4 defined the weighted, fractional maximal function $M^{N-\alpha}_{w,\rho}\mu$ of a measure $\mu \in \mathcal{M}(\mathbf{R}^N)$ by

$$M^{N-\alpha}_{w,\rho}\mu(x) = \sup_{0 < r \leq \rho} \frac{|\mu|(B_r(x))}{h^{N-\alpha}_w(B_r(x))}, \quad x \in \mathbf{R}^N,$$

where, as before, $h^{N-\alpha}_w(B_r(x)) = r^{-\alpha}w(B_r(x))$. We now define a weighted Morrey space.[†]

[†]Morrey spaces were introduced by C. B. Morrey Jr. in [83]. For a comprehensive survey of these spaces and historical remarks, see Giaquinta [46, pp. 64–74].

Definition 4.1.7. Let $0 < \alpha < N$ and $0 < \rho \leq \infty$. Let w be a weight. The *Morrey space* $L_{w,\rho}^{1,N-\alpha}(\mathbf{R}^N)$ consists of all measures $\mu \in \mathcal{M}(\mathbf{R}^N)$ with finite norm

$$\|\mu\|_{L_{w,\rho}^{1,N-\alpha}(\mathbf{R}^N)} = \|M_{w,\rho}^{N-\alpha}\mu\|_{L^\infty(\mathbf{R}^N)}.$$

The proof of the existence theorem below is a variation of a proof by A. Carlsson [20, pp. 12–13].[†] A minor difference is the fact that Carlsson's result concerns $\mathcal{H}_\infty^{N-m}$-quasicontinuous representatives.

Theorem 4.1.8. *Let $w \in A_1$, and let m be an integer, $1 \leq m < N$. Then every function in $W_w^{m,1}(\mathbf{R}^N)$ has a Borel measurable, $\mathcal{H}_{w,1}^{N-m}$-quasicontinuous representative.*

Proof. Let $u \in W_w^{m,1}(\mathbf{R}^N)$ and choose a sequence $\{u_n\}_{n=1}^\infty$ of functions in $C_0^\infty(\mathbf{R}^N)$ such that u_n converges to u in $W_w^{m,1}(\mathbf{R}^N)$ so fast that

$$\|u_{n+1} - u_n\|_{W_w^{m,1}(\mathbf{R}^N)} \leq 4^{-n}$$

for $n = 1, 2, \dots$. Note that the series

$$v(x) = \sum_{i=1}^\infty |u_{i+1}(x) - u_i(x)|$$

is defined and lower semicontinuous for every $x \in \mathbf{R}^N$. We now claim that v is finite $\mathcal{H}_{w,1}^{N-m}$-q.e. Let B be the set of $x \in \mathbf{R}^N$ for which $v(x) = \infty$. Suppose that $\mathcal{H}_{w,1}^{N-m}(B) > 0$. Then, by Frostman's lemma (Theorem 3.4.27), there is a measure $\mu \in \mathcal{M}^+(B)$ such that $\mu(B) > 0$ and $\|\mu\|_{L_{w,1}^{1,N-m}(\mathbf{R}^N)} \leq 1$. By appealing to Theorem 2.6.4, we then get

$$\int_B v \, d\mu = \sum_{i=1}^\infty \int_B |u_{i+1} - u_i| \, d\mu$$

$$\leq C\|\mu\|_{L_{w,1}^{1,N-m}(\mathbf{R}^N)} \sum_{i=1}^\infty \|u_{i+1} - u_i\|_{W_w^{m,1}(\mathbf{R}^N)} \leq C,$$

which is a contradiction because $\int_B v \, d\mu = \infty$. This proves the claim. It follows that the series

$$\tilde{u}(x) = \sum_{i=1}^\infty (u_{i+1}(x) - u_i(x))$$

is absolutely convergent $\mathcal{H}_{w,1}^{N-m}$-q.e. But since we also have

$$u_n(x) = u_1(x) + \sum_{i=1}^{n-1}(u_{i+1}(x) - u_i(x)),$$

[†]This result is implicitly contained in earlier results by D. R. Adams [4, p. 122].

we see that $\tilde{u}(x) = \lim_{n\to\infty} u_n(x)$ for $x \in \mathbf{R}^N \setminus B$. The function $\tilde{u}$ is of course Borel measurable, being the limit of smooth functions.

It remains to show that $\tilde{u}$ is an $\mathcal{H}^{N-m}_{w,1}$-quasicontinuous representative of u. For $n = 1, 2, \ldots$, we define

$$E_n = \{x \in \mathbf{R}^N \setminus B \; ; \; |\tilde{u}(x) - u_n(x)| \geq 2^{-n}\}.$$

If $\mu \in \mathcal{M}^+(B)$ and $\|\mu\|_{L^{1,N-m}_{w,1}(\mathbf{R}^N)} \leq 1$, then using the fact that

$$\tilde{u}(x) - u_n(x) = \sum_{i=n}^{\infty} (u_{i+1}(x) - u_i(x))$$

for $x \in \mathbf{R}^N \setminus B$, we obtain as above

$$\mu(E_n) \leq 2^n \int_{E_n} |\tilde{u} - u_n| \, d\mu \leq C2^n \sum_{i=n}^{\infty} \|u_{i+1} - u_i\|_{W^{m,1}_w(\mathbf{R}^N)} \leq C2^{-n}.$$

From this inequality, we infer that $\mathcal{H}^{N-m}_{w,1}(E_n) \leq C2^{-n}$. Now let $F_j = \bigcup_{n=j}^{\infty} E_n$ and $F = \bigcap_{j=1}^{\infty} F_j$. Then, by Proposition 3.4.3(c), $\mathcal{H}^{N-m}_{w,1}(F_j) \leq C2^{-j}$, and thus, $\mathcal{H}^{N-m}_{w,1}(F) = 0$ since $F \subset F_j$ for every j. For $x \in (F \cup B)^c$, we have $u_n(x) \to \tilde{u}(x)$. Hence, by Remark 3.4.17, $u_n \to \tilde{u}$ a.e. Since we may assume that $u_n \to u$ a.e., we conclude that $\tilde{u}(x) = u(x)$ for a.e. $x \in \mathbf{R}^N$, i.e., $\tilde{u}$ is a representative of u. Finally, note that for every $\varepsilon > 0$, we can choose F_j and an open set $G \supset F_j \cup B$ such that $\mathcal{H}^{N-m}_{w,1}(G) < \varepsilon$. But $\tilde{u}$ is continuous on G^c because $u_n \to \tilde{u}$ uniformly outside $F_j \cup B$. $\square$

With $w = 1$, the uniqueness theorem below was proved by Carlsson in [20, pp. 13–15]. Carlsson's proof builds on ideas from Wallin [106, pp. 72–73]. Our proof follows the proof in [20] closely.

Theorem 4.1.9. *Let $w \in A_1$, and let $0 < \alpha < N$. Let O be an open subset of $\mathbf{R}^N$ for which there exists two positive constants a and R such that*

$$\frac{|O \cap B_r(x)|}{|B_r(x)|} \geq a$$

for every $x \in \bar{O}$ and $0 < r \leq R$. Suppose that u_1 and u_2 are two Borel measurable, $\mathcal{H}^{N-\alpha}_{w,1}$-quasicontinuous functions on $\mathbf{R}^N$ so that $u_1 = u_2$ a.e. on O. Then $u_1 = u_2$ $\mathcal{H}^{N-\alpha}_{w,1}$-q.e. on $\bar{O}$.

Proof. Set $u = u_1 - u_2$, and define $E = \{x \in \bar{O} \; ; \; u(x) \neq 0\}$. Since it suffices to show that $\mathcal{H}^{N-\alpha}_{w,1}(E \cap B) = 0$ for every open ball B, we can assume that E is bounded. By Theorem 3.4.6, the capacity of E is then finite. Suppose that $\mathcal{H}^{N-\alpha}_{w,1}(E) = A > 0$. Let ε, $0 < \varepsilon < A$, be a constant as yet to be specified, and let G be an open set with $\mathcal{H}_{w,1}^{N-\alpha}(G) < \varepsilon$ such that $u|_{G^c}$ is continuous. We then have $\mathcal{H}_{w,1}^{N-\alpha}(E \setminus G) > A - \varepsilon$. Using the capacitability theorem (Theorem 3.4.25), we can find a compact subset K of $E \setminus G$ such that $\mathcal{H}_{w,1}^{N-\alpha}(K) > A - \varepsilon$. For $n = 2, 3, \ldots$, let

$$K_n = \{x \in \mathbf{R}^N ; \operatorname{dist}(x, K) \leq 1/n\}.$$

Frostman's lemma now implies the existence of measures $\mu_n \in \mathcal{M}^+(K_n \cap \bar{O})$, satisfying $\|\mu_n\|_{L_{w,2}^{1,N-\alpha}(\mathbf{R}^N)} \leq 1$ and $\mu_n(K_n \cap \bar{O}) \geq C\mathcal{H}_{w,2}^{N-\alpha}(K_n \cap \bar{O})$. Let $\varphi \in C_0^\infty(B_1(0))$ be such that $0 \leq \varphi \leq 1$, $\varphi \geq \frac{1}{2}$ on $B_{1/2}(0)$, and $\int_{\mathbf{R}^N} \varphi \, dx = 1$. Define $\varphi_n(x) = n^N \varphi(nx)$, and let ν_n be defined by $d\nu_n(x) = (\varphi_n * \mu_n) \, dx$.

We now show that $\|\nu_n\|_{L_{w,1}^{1,N-\alpha}(\mathbf{R}^N)} \leq C$. First of all,

$$\varphi_n * \mu_n(y) = n^N \int_{|y-z|<1/n} \varphi(n(y-z)) \, d\mu_n(z) \leq n^N \mu_n(B_{1/n}(y))$$
$$\leq n^N h_w^{N-\alpha}(B_{1/n}(y)).$$

Let $x \in \mathbf{R}^N$ be arbitrary. If $0 < r < 1/n$, then

$$\nu_n(B_r(x)) = \int_{B_r(x)} (\varphi_n * \mu_n) \, dy \leq n^N \int_{B_r(x)} h_w^{N-\alpha}(B_{1/n}(y)) \, dy$$
$$\leq 2^\alpha n^N |B_r(x)| h_w^{N-\alpha}(B_{2/n}(x)).$$

The strong doubling property of w now shows that

$$h_w^{N-\alpha}(B_{2/n}(x)) = \left(\frac{n}{2}\right)^\alpha w(B_{2/n}(x)) \leq C\left(\frac{n}{2}\right)^\alpha \frac{|B_{2/n}(x)|}{|B_r(x)|} w(B_r(x))$$
$$= C(rn)^\alpha n^{-N} |B_r(x)|^{-1} h_w^{N-\alpha}(B_r(x))$$
$$\leq Cn^{-N} |B_r(x)|^{-1} h_w^{N-\alpha}(B_r(x)),$$

which implies that

$$\frac{\nu_n(B_r(x))}{h_w^{N-\alpha}(B_r(x))} \leq C. \tag{4.1.3}$$

If on the other hand $1/n \leq r < 1$, we have

$$\nu_n(B_r(x)) = \int_{\mathbf{R}^N} \left(\int_{|x-y|<r} \varphi_n(y-z) \, dy \right) d\mu_n(z)$$
$$\leq \int_{\mathbf{R}^N} \chi_{B_{r+1/n}(x)}(z) \, d\mu_n(z)$$
$$= \mu_n(B_{r+1/n}(x)) \leq h_w^{N-\alpha}(B_{r+1/n}(x))$$
$$\leq Ch_w^{N-\alpha}(B_r(x)),$$

where the last inequality again follows from the strong doubling property of w, so (4.1.3) holds in this case as well. We have thus proved that $\|\nu_n\|_{L^{1,N-\alpha}_{w,1}(\mathbf{R}^N)} \leq C$.

Let K_n^* be the support of ν_n. Since $\|\nu_n\|_{L^{1,N-\alpha}_{w,1}(\mathbf{R}^N)} \leq C$ and $|E \cap O| = 0$, we have

$$\mathcal{H}^{N-\alpha}_{w,1}(K_n^* \cap O \cap E^c) \geq C\nu_n(O \cap E^c) = C\nu_n(O).$$

Moreover, if n is so large that $\frac{1}{2n} \leq R$, then

$$\nu_n(O) = n^N \int_{\mathbf{R}^N} \left(\int_{O \cap B_{1/n}(z)} \varphi(n(y-z))\,dy \right) d\mu_n(z)$$

$$\geq C2^{-(N+1)} \int_{\mathbf{R}^N} \frac{|O \cap B_{1/2n}(z)|}{|B_{1/2n}(z)|}\,d\mu_n(z)$$

$$\geq 2^{-(N+1)} a\mu_n(K_n \cap \bar{O}) \geq C\mathcal{H}^{N-\alpha}_{w,2}(K_n \cap \bar{O}).$$

By Lemma 3.4.15, $\mathcal{H}^{N-\alpha}_{w,2}(K_n \cap \bar{O}) \geq C\mathcal{H}^{N-\alpha}_{w,1}(K_n \cap \bar{O})$. Because $K \subset K_n$ and $K \subset E \setminus G \subset \bar{O}$, we find

$$\mathcal{H}^{N-\alpha}_{w,1}(K_n \cap \bar{O}) \geq \mathcal{H}^{N-\alpha}_{w,1}(K) > A - \varepsilon,$$

and hence, $\mathcal{H}^{N-\alpha}_{w,1}(K_n^* \cap O \cap E^c) > C(A - \varepsilon)$. If we now choose ε so small that $C(A - \varepsilon) > \varepsilon$, we obtain $\mathcal{H}^{N-\alpha}_{w,1}(K_n^* \cap O \cap E^c) > \mathcal{H}^{N-\alpha}_{w,1}(G)$. This implies that there are points $x_n \in K_n^* \cap O \cap E^c \cap G^c$ for every $n \geq \frac{1}{2R}$. Note that $K_n^* \subset K_{n/2}$, so by compactness, we may assume that x_n converges to some point x_0. There are also $y_n \in K$ such that $|x_n - y_n| \leq 2/n$, so it follows that $y_n \to x_0$, and hence that $x_0 \in K$. Note that u is continuous at x_0 and $u(x_0) \neq 0$ because $K \subset E \setminus G$. But $0 = u(x_n) \to u(x_0)$ since $x_n \in (E \cup G)^c$. We have thus reached a contradiction, and conclude that $\mathcal{H}^{N-\alpha}_{w,1}(E) = 0$. $\square$

4.2. Measures in the dual of $W^{m,p}_w(\Omega)$

The purpose of this short section is to give sufficient conditions for a positive Radon measure to belong to the dual $W^{m,p}_w(\Omega)^*$ of $W^{m,p}_w(\Omega)$, if the action of the measure on $W^{m,p}_w(\Omega)$ is given through integration. These will be used in Section 4.3 to prove some Poincaré type inequalities. Our results extend earlier results by A. Carlsson for the non-weighted case [20, p. 15].

4.2.1. The case $1 < p < \infty$

Theorem 4.2.1. *Let m be an integer, $1 \leq m < N$, and let $1 < p < \infty$. Let w be an A_p weight. Suppose that Ω is a bounded Lipschitz domain in $\mathbf{R}^N$. Let μ be a positive Radon measure, supported in $\bar{\Omega}$, such that $\mathcal{G}_m\mu \in L^{p'}_{w'}(\mathbf{R}^N)$. Then*

μ belongs to the dual of $W_w^{m,p}(\Omega)$, acting on $W_w^{m,p}(\Omega)$ through the pairing

$$\langle u, \mu \rangle = \int_{\mathbf{R}^N} \widetilde{Eu}\, d\mu \qquad (4.2.1)$$

for $u \in W_w^{m,p}(\Omega)$. Here, $E : W_w^{m,p}(\Omega) \to W_w^{m,p}(\mathbf{R}^N)$ is any bounded extension operator and $\widetilde{Eu}$ is any Borel measurable, $B_{m,p}^w$-quasicontinuous representative of Eu. Moreover,

$$\|\mu\|_{W_w^{m,p}(\Omega)^*} \leq C\|\mathcal{G}_m\mu\|_{L_{w'}^{p'}(\mathbf{R}^N)}, \qquad (4.2.2)$$

where the constant C only depends on m, N, p, and the A_p constant of w.

Remark 4.2.2. In the case $\Omega = \mathbf{R}^N$, the condition $\mathcal{G}_m\mu \in L_{w'}^{p'}(\mathbf{R}^N)$ is also necessary for μ to belong to the dual of $W_w^{m,p}(\mathbf{R}^N)$. This can be seen as in Ziemer [108, p. 205].

Proof. Suppose that $u \in W_w^{m,p}(\Omega)$, and let E be the extension operator in Theorem 2.1.13. Then, by Theorem 3.3.2, $Eu = \mathcal{G}_m f$ for some $f \in L_w^p(\mathbf{R}^N)$. We also know from the last section that $\mathcal{G}_m f$ is $B_{m,p}^w$-quasicontinuous. Fubini's theorem and Hölder's inequality then show that

$$\left| \int_{\mathbf{R}^N} \mathcal{G}_m f\, d\mu \right| = \left| \int_{\mathbf{R}^N} (\mathcal{G}_m\mu) f\, dx \right| \leq \|\mathcal{G}_m\mu\|_{L_{w'}^{p'}(\mathbf{R}^N)} \|f\|_{L_w^p(\mathbf{R}^N)}$$
$$\leq C\|\mathcal{G}_m\mu\|_{L_{w'}^{p'}(\mathbf{R}^N)} \|u\|_{W_w^{m,p}(\Omega)},$$

i.e., the inequality (4.2.2) holds.

We now proceed to show that the action of μ on $W_w^{m,p}(\Omega)$ is well-defined in the sense that the right-hand side in (4.2.1) does not depend on the various choices involved. First, if $\widehat{Eu}$ is another $B_{m,p}^w$-quasicontinuous representative of Eu, it follows from the uniqueness theorem that $\widehat{Eu} = \mathcal{G}_m f$ $B_{m,p}^w$-q.e. By the dual definition in Proposition 3.2.3, $\widehat{Eu} = \mathcal{G}_m f$ μ-a.e., so $\langle u, \mu \rangle$ is independent of the particular $B_{m,p}^w$-quasicontinuous representative we choose. Now suppose that $E_1 : W_w^{m,p}(\Omega) \to W_w^{m,p}(\mathbf{R}^N)$ is another extension operator. Then

$$\mathcal{G}_m f = Eu = u = E_1 u = \widetilde{E_1 u} \quad \text{a.e. on } \Omega.$$

This implies that $\widetilde{E_1 u} = \mathcal{G}_m f$ μ-a.e. on $\overline{\Omega}$. In view of the fact that supp $\mu \subset \overline{\Omega}$, we obtain

$$\int_{\mathbf{R}^N} \widetilde{E_1 u}\, d\mu = \int_{\mathbf{R}^N} \mathcal{G}_m f\, d\mu. \qquad \square$$

4.2.2. The case $p = 1$

To be able to prove a result corresponding to Theorem 4.2.1 for $p = 1$, we first extend Theorem 2.6.4 to functions belonging to $W_w^{m,1}(\mathbf{R}^N)$.

Theorem 4.2.3. *Let m be an integer, $1 \le m < N$, and let w be an A_1 weight. Then there is a constant C, depending only on m, N, and the A_1 constant of w, such that if μ is a positive Radon measure in $L_{w,1}^{1,N-m}(\mathbf{R}^N)$, then the inequality*

$$\int_{\mathbf{R}^N} |\tilde{u}| \, d\mu \le C \|\mu\|_{L_{w,1}^{1,N-m}(\mathbf{R}^N)} \|u\|_{W_w^{m,1}(\mathbf{R}^N)} \qquad (4.2.3)$$

holds for every $u \in W_w^{m,1}(\mathbf{R}^N)$, where $\tilde{u}$ is any Borel measurable, $\mathcal{H}_{w,1}^{N-m}$-quasicontinuous representative of u.

Proof. Let $u \in W_w^{m,1}(\mathbf{R}^N)$, and let $\tilde{u}$ be the $\mathcal{H}_{w,1}^{N-m}$-quasicontinuous representative of u in Theorem 4.1.8. Then $\tilde{u}$ is defined $\mathcal{H}_{w,1}^{N-m}$-q.e. and hence, by Frostman's lemma, μ-a.e. By the proof of Theorem 4.1.8, $\tilde{u} = \lim_{n\to\infty} u_n$ μ-a.e., where u_n are smooth functions. The inequality (4.2.3) now follows from Fatou's lemma and Theorem 2.6.4. If $\bar{u}$ is another Borel measurable, $\mathcal{H}_{w,1}^{N-m}$-quasicontinuous representative of u, then, according to Frostman's lemma and the uniqueness theorem (Theorem 4.1.9), $\bar{u} = \tilde{u}$ μ-a.e., which shows that the left-hand side in (4.2.3) does not depend on the specific quasicontinuous representative chosen. $\square$

Theorem 4.2.4. *Let m be an integer, $1 \le m < N$, and let w be an A_1 weight. Suppose that Ω is a bounded Lipschitz domain in $\mathbf{R}^N$. Let μ be a positive Radon measure, belonging to $L_{w,1}^{1,N-m}(\mathbf{R}^N)$, with support in $\overline{\Omega}$. Then μ is in the dual of $W_w^{m,1}(\Omega)$, acting on $W_w^{m,1}(\Omega)$ through the pairing*

$$\langle u, \mu \rangle = \int_{\mathbf{R}^N} \widetilde{Eu} \, d\mu \qquad (4.2.4)$$

for $u \in W_w^{m,1}(\Omega)$, where $E : W_w^{m,1}(\Omega) \to W_w^{m,1}(\mathbf{R}^N)$ is any bounded extension operator and $\widetilde{Eu}$ is any Borel measurable, $\mathcal{H}_{w,1}^{N-m}$-quasicontinuous representative of Eu. Moreover,

$$\|\mu\|_{W_w^{m,1}(\Omega)^*} \le C \|\mu\|_{L_{w,1}^{1,N-m}(\mathbf{R}^N)}, \qquad (4.2.5)$$

where the constant C only depends on m, N, and the A_1 constant of w.

Remark 4.2.5. In the case $\Omega = \mathbf{R}^N$, it follows directly from Theorem 2.6.4 that the condition $\mu \in L_{w,1}^{1,N-m}(\mathbf{R}^N)$ is necessary for μ to belong to $W_w^{m,1}(\mathbf{R}^N)^*$.

Proof. The proof is essentially the same as that of Theorem 4.2.1. The inequality (4.2.5) is a consequence of (4.2.3). It has already been shown that the right-hand side of (4.2.4) is independent of which quasicontinuous representative of u we choose. The independence of the particular extension operator is proved as in Theorem 4.2.1, with the only difference that we use Frostman's lemma instead of the dual definition of Bessel capacity. $\square$

4.3. Poincaré type inequalities

In this section, several Poincaré type inequalities are obtained by means of an approach by N. G. Meyers [79]. These kind of inequalities were first considered by V. G. Maz'ya in [71], [74], and [76]. Meyers' methods have later been developed by A. Carlsson [20] and W. P. Ziemer [108].[†] Another approach for proving Poincaré type inequalities is due to L. I. Hedberg [56], [57]. More information and further references can be found in the monographs by Adams–Hedberg [7], Maz'ya [76], and Ziemer [108].

For $1 < p < \infty$, similar weighted Poincaré type inequalities have earlier been obtained by N. O. Belova [12], [13], [14].

4.3.1. The case $1 < p < \infty$

Lemma 4.3.1. *Let k and m be integers, $0 \leq k \leq m - 1$, and let $1 < p < \infty$ and $w \in A_p$. Suppose that μ_α, $|\alpha| \leq k$, are positive Radon measures, supported in $\overline{B_1(0)}$, such that $\mathcal{G}_{m-|\alpha|}\mu_\alpha \in L^{p'}_{w'}(\mathbf{R}^N)$, where $w' = w^{-1/(p-1)}$, and $\mu_\alpha(\overline{B_1(0)}) = 1$. If $k \leq m - 2$, let $\omega \in C_0^\infty(B_1(0))$ satisfy $\int_{B_1(0)} \omega \, dx = 1$, and define $d\mu_\alpha = \omega \, dx$ for $k + 1 \leq |\alpha| \leq m - 1$. Then there exists a projection $L : W_w^{m,p}(B_1(0)) \to \mathcal{P}_{m-1}$ such that if $u \in W_w^{m,p}(B_1(0))$, then*

$$\|Lu\|_{L_w^p(B_1(0))} \leq C\left(w(B_1(0))^{1/p}\sum_{0 \leq |\alpha| \leq k} \|\mathcal{G}_{m-|\alpha|}\mu_\alpha\|_{L^{p'}_{w'}(\mathbf{R}^N)} + C_k\right)\|u\|_{W_w^{m,p}(B_1(0))},$$

(4.3.1)

where $C_k = 1$, if $k \leq m - 2$, and $C_k = 0$, if $k = m - 1$. The constant C depends only on k, m, N, p, and the A_p constant of w. The projection L is uniquely determined by the requirement that

$$\int_{\overline{B_1(0)}} D^\alpha u \, d\mu_\alpha = \int_{\overline{B_1(0)}} D^\alpha(Lu) \, d\mu_\alpha, \quad |\alpha| \leq m - 1. \qquad (4.3.2)$$

Proof. If we for $u \in W_w^{m,p}(B_1(0))$ let $Lu(x) = \sum_{|\beta| \leq m-1} a_\beta x^\beta$, then (4.3.2) yields

$$a_\beta = \frac{1}{\beta!} \int_{\overline{B_1(0)}} D^\beta u \, d\mu_\beta, \quad |\beta| = m - 1,$$

while we for $0 \leq |\beta| \leq m - 2$ have the recursion formula

$$a_\beta = \frac{1}{\beta!} \int_{\overline{B_1(0)}} D^\beta u \, d\mu_\beta - \sum_{1 \leq |\gamma| \leq m-1-|\beta|} a_{\beta+\gamma} \binom{\beta+\gamma}{\gamma} \int_{\overline{B_1(0)}} x^\gamma \, d\mu_\beta. \quad (4.3.3)$$

[†]It should be pointed out that we do not use Meyers' functional analytic lemma, but rather the proof of the lemma. The reason for this is that if we used this lemma, we would obtain inequalities in which the dependence of the constant on the weight could not be determined explicitly, and therefore would be of little use.

This shows that L exists and is uniquely determined. Since the operator L is unique, it has to be a projection. Evidently,

$$\|Lu\|_{W_w^{m,p}(B_1(0))} \leq C w(B_1(0))^{1/p} \sum_{0 \leq |\beta| \leq m-1} |a_\beta|.$$

Suppose that $k \leq m - 2$. If $0 \leq |\beta| \leq k$, then according to (4.3.3) and Theorem 4.2.1,

$$|a_\beta| \leq C \|\mathcal{G}_{m-|\beta|}\mu_\beta\|_{L_{w'}^{p'}(\mathbf{R}^N)} \|D^\beta u\|_{W_w^{m-|\beta|,p}(B_1(0))} + C \sum_{1 \leq |\gamma| \leq m-1-|\beta|} |a_{\beta+\gamma}|,$$

and if $k + 1 \leq |\beta| \leq m - 1$, then by Remark 1.2.4.1,

$$|a_\beta| \leq C \int_{B_1(0)} |D^\beta u| \omega \, dx + C \sum_{1 \leq |\gamma| \leq m-1-|\beta|} |a_{\beta+\gamma}|$$

$$\leq C w(B_1(0))^{-1/p} \left(\int_{B_1(0)} |D^\beta u|^p w \, dx \right)^{1/p} + C \sum_{1 \leq |\gamma| \leq m-1-|\beta|} |a_{\beta+\gamma}|.$$

The lemma now follows by induction.

If $k = m - 1$, the proof is almost the same with the only difference that the last step corresponding to $k + 1 \leq |\beta| \leq m - 1$ is unnecessary.   □

The following convention will be used below. Let $u \in W_w^{m,p}(\mathbf{R}^N)$, where $w \in A_p$, and let $E \subset \mathbf{R}^N$ be arbitrary. If α is a multi-index, $0 \leq |\alpha| \leq m$, then by $D^\alpha u|_E = 0$ we shall mean that $D^\alpha u$ has an $R_{m-|\alpha|,p;1}^w$-quasicontinuous representative $\widetilde{D^\alpha u}$ such that $\widetilde{D^\alpha u} = 0$ $R_{m-|\alpha|,p;1}^w$-q.e. on E.

Lemma 4.3.2. *Let k and m be integers, $0 \leq k \leq m-1$, and let $1 < p < \infty$ and $w \in A_p$. Let $u \in W_w^{m,p}(B_1(0))$, and suppose that $D^\alpha u|_E = 0$ for $0 \leq |\alpha| \leq k$, where $E \subset B_1(0)$ is a Suslin set with $R_{m-k,p;1}^w(E) > 0$. Then there exists a polynomial $P \in \mathcal{P}_{m-1}$, $P(x) = \sum_{|\beta| \leq m-1} a_\beta x^\beta$, such that*

$$\|u - Lu\|_{W_w^{m,p}(B_1(0))} \leq C \frac{w(B_1(0))^{1/p}}{R_{m-k,p;1}^w(E)^{1/p}} \|\nabla^m u\|_{L_w^p(B_1(0))} \tag{4.3.4}$$

and

$$|a_\beta| \leq C w(B_1(0))^{-1/p} \|\nabla^{k+1} u\|_{L_w^p(B_1(0))}, \quad 0 \leq |\beta| \leq m - 1. \tag{4.3.5}$$

In the case $k = m - 1$, P is identically 0. The constant C depends only on k, m, N, p and the A_p constant of w.

Proof. It follows from the dual definition of the Bessel capacity (Proposition 3.2.3) that there are measures $\mu_\alpha \in \mathcal{M}^+(E)$, $0 \le |\alpha| \le k$, such that $\|\mathcal{G}_{m-|\alpha|}\mu_\alpha\|_{L^{p'}_{w'}(\mathbf{R}^N)} \le 1$ and

$$\mu_\alpha(E) \ge \tfrac{1}{2} B^w_{m-|\alpha|,p}(E)^{1/p}.$$

Using Theorem 3.3.7 and the fact that $R^w_{m-|\alpha|,p;1}(E) \ge R^w_{m-k,p;1}(E)$, we then get

$$\mu_\alpha(E) \ge \tfrac{1}{2} B^w_{m-k,p}(E)^{1/p} \ge C R^w_{m-k,p;1}(E)^{1/p}. \tag{4.3.6}$$

If $k \le m - 2$, we let μ_α, $k + 1 \le |\alpha| \le m - 1$ be as in Lemma 4.3.1. Let $u \in W^{m,p}_w(B_1(0))$. Let $P = Lu$ be the polynomial in Lemma 4.3.1, and let $L' : W^{m,p}_w(B_1(0)) \to \mathcal{P}_{m-1}$ be the projection in Corollary 2.1.15. We then have

$$\|u - L'u\|_{W^{m,p}_w(B_1(0))} \le C\|\nabla^m u\|_{L^p_w(B_1(0))}. \tag{4.3.7}$$

By writing $u - Lu = u - L'u - L(u - L'u)$, it follows that

$$\|u - Lu\|_{W^{m,p}_w(B_1(0))} \le C\|\nabla^m u\|_{L^p_w(B_1(0))} + \|L(u - L'u)\|_{W^{m,p}_w(B_1(0))}.$$

Lemma 4.3.1 and (4.3.7) now imply that

$$\|L(u - L'u)\|_{W^{m,p}_w(B_1(0))}$$
$$\le C\left(w(B_1(0))^{1/p} \sum_{0 \le |\alpha| \le k} \frac{\|\mathcal{G}_{m-|\alpha|}\mu_\alpha\|_{L^{p'}_{w'}(\mathbf{R}^N)}}{\mu_\alpha(\overline{B_1(0)})} + C_k \right)\|\nabla^m u\|_{L^p_w(B_1(0))}.$$

We then use (4.3.6) together with the fact that

$$R^w_{m-k,p;1}(E) \le R^w_{m-k,p;1}(\overline{B_1(0)}) \le Cw(B_1(0)),$$

(see Lemma 3.3.12), to obtain

$$\|L(u - L'u)\|_{W^{m,p}_w(B_1(0))} \le C \frac{w(B_1(0))^{1/p}}{R^w_{m-k,p;1}(E)^{1/p}}\|\nabla^m u\|_{L^p_w(B_1(0))},$$

which proves the inequality (4.3.4).

It remains to estimate the coefficients of P. First suppose that $k \le m - 2$. Let $k + 1 \le |\beta| \le m - 1$ and $0 \le \gamma \le \beta$. Then

$$\int_{\overline{B_1(0)}} D^\beta u \, d\mu_\beta = \int_{B_1(0)} D^\beta u \, \omega \, dx = (-1)^{|\gamma|} \int_{B_1(0)} D^\gamma u D^{\beta-\gamma}\omega \, dx.$$

Thus, if we choose γ such that $|\gamma| = k + 1$, this identity together with the formula (4.3.3) and the A_p condition show that (4.3.5) holds. If $0 \le |\beta| \le k$, then since $D^\beta u|_E = 0$ and μ_β is concentrated to E,

$$|a_\beta| \le C \sum_{1 \le |\gamma| \le m-1-|\beta|} |a_{\beta+\gamma}| = C \sum_{|\beta|+1 \le |\gamma| \le m-1} |a_\gamma|.$$

Hence, (4.3.5) holds for $0 \leq |\beta| \leq m - 1$.

If $k = m - 1$, the polynomial P is identically 0 and there is nothing to prove. $\square$

Lemma 4.3.3. *Let $0 < \alpha < N$ and $1 < p < \infty$. Let w be an A_p weight. Let $a \in \mathbf{R}^N$ and $0 < \rho < \infty$. For $E \subset \mathbf{R}^N$, set*

$$(E - a)/\rho = \{(x - a)/\rho \; ; \; x \in E\}.$$

Set $\tilde{w}(x) = w(\rho x + a)$. Then

$$R^w_{\alpha,p;\rho}(E) = \rho^{N-\alpha p} R^{\tilde{w}}_{\alpha,p;1}((E - a)/\rho). \tag{4.3.8}$$

Proof. Let f be a nonnegative function such that $\mathcal{I}_{\alpha,1}f((x - a)/\rho) \geq 1$ for $x \in E$ and $\|f\|^p_{L^p_{\tilde{w}}(\mathbf{R}^N)} < R^{\tilde{w}}_{\alpha,p;1}((E - a)/\rho) + \varepsilon$, where $\varepsilon > 0$ is arbitrary. It is easy to see that

$$\mathcal{I}_{\alpha,1}f((x - a)/\rho) = \mathcal{I}_{\alpha,\rho}f_\rho(x),$$

where $f_\rho(x) = \rho^{-\alpha}f((x - a)/\rho)$. We thus have $\mathcal{I}_{\alpha,\rho}f_\rho \geq 1$ on E. Furthermore,

$$\begin{aligned} R^w_{\alpha,p;\rho}(E) &\leq \int_{\mathbf{R}^N} f^p_\rho w \, dx = \rho^{N-\alpha p} \int_{\mathbf{R}^N} f^p \tilde{w} \, dx \\ &< \rho^{N-\alpha p}(R^{\tilde{w}}_{\alpha,p;1}((E - a)/\rho) + \varepsilon). \end{aligned}$$

This proves that

$$R^w_{\alpha,p;\rho}(E) \leq \rho^{N-\alpha p} R^{\tilde{w}}_{\alpha,p;1}((E - a)/\rho).$$

The reverse inequality is proved in the same manner. $\square$

Using Lemma 4.3.3, we obtain the following corollaries from Lemma 4.3.2.

Corollary 4.3.4. *Let m be an integer, $1 < p < \infty$ and $w \in A_p$. Then there is a constant C, depending only on m, N, p and the A_p constant of w, such that if $u \in W^{m,p}_w(B_r(a))$ and $D^\alpha u|_E = 0$ for $|\alpha| \leq m - 1$, where $E \subset B_r(a)$ is a Suslin set with $R^w_{1,p;r}(E) > 0$, then*

$$\int_{B_r(a)} |u|^p w \, dx \leq C \frac{w(B_r(a))}{r^p R^w_{1,p;r}(E)} r^{mp} \int_{B_r(a)} |\nabla^m u|^p w \, dx. \tag{4.3.9}$$

Corollary 4.3.5. *Let k and m be integers, $0 \leq k \leq m - 1$, and let $1 < p < \infty$ and $w \in A_p$. Then there is a constant C, depending only on k, m, N, p and the A_p constant of w, such that if $u \in W^{m,p}_w(B_r(a))$ and $D^\alpha u|_E = 0$ for $0 \leq |\alpha| \leq k$, where $E \subset B_r(a)$ is a Suslin set with $R^w_{m-k,p;r}(E) > 0$, then*

$$\begin{aligned} \int_{B_r(a)} |u|^p w \, dx \leq{}& C r^{(k+1)p} \int_{B_r(a)} |\nabla^{k+1} u|^p w \, dx \\ &+ C \frac{r^{-(m-k)p} w(B_r(a))}{R^w_{m-k,p;r}(E)} r^{mp} \int_{B_r(a)} |\nabla^m u|^p w \, dx. \end{aligned}$$

4.3.2. The case $p = 1$

This section contains counterparts to the results obtained in the previous section. Since the proofs are completely analogous, we shall be content with stating our results.

We begin by proving two lemmas.

Lemma 4.3.6. *Let $0 \leq \alpha \leq \beta < N$ and let w be a weight. Then*

$$\mathcal{H}_{w,1}^{N-\beta}(E) \geq \mathcal{H}_{w,1}^{N-\alpha}(E)$$

for every subset E of $\mathbf{R}^N$.

Proof. If $r \leq 1$, then $r^\beta \leq r^\alpha$, so $h_w^{N-\beta}(B_r(x)) \geq h_w^{N-\alpha}(B_r(x))$ for every $x \in \mathbf{R}^N$. $\square$

Lemma 4.3.7. *Let $0 \leq \alpha < N$ and let w be a weight. Let $a \in \mathbf{R}^N$ and $0 < \rho < \infty$. With the same notation as in Lemma 4.3.3, for every set $E \subset \mathbf{R}^N$,*

$$\mathcal{H}_{w,\rho}^{N-\alpha}(E) = \mathcal{H}_{\tilde{w},1}^{N-\alpha}((E-a)/\rho). \tag{4.3.10}$$

Proof. The equality (4.3.10) is a consequence of the following simple facts:

1. $h_w^{N-\alpha}(B_r(a)) = \rho^{N-\alpha} h_{\tilde{w}}^{N-\alpha}(B_{r/\rho}((x-a)/\rho))$.

2. There is a one-to-one correspondence between admissible coverings of E and $(E-a)/\rho$ given by $B_r(x) \leftrightarrow B_{r/\rho}((x-a)/\rho)$. $\square$

Lemma 4.3.8. *Let k and m be integers, $0 \leq k \leq m-1$, and let $w \in A_1$. Suppose that μ_α, $|\alpha| \leq k$, are positive Radon measures, supported in $\overline{B_1(0)}$, such that $\mu_\alpha \in L_{w,1}^{1,N-(m-\alpha)}(\mathbf{R}^N)$ and $\mu_\alpha(\overline{B_1(0)}) = 1$. If $k \leq m-2$, let the function $\omega \in C_0^\infty(B_1(0))$ satisfy $\int_{B_1(0)} \omega \, dx = 1$, and define $d\mu_\alpha = \omega \, dx$ for $k+1 \leq |\alpha| \leq m-1$. Then there exists a projection $L : W_w^{m,1}(B_1(0)) \to \mathcal{P}_{m-1}$ such that if $u \in W_w^{m,1}(B_1(0))$, then*

$$\|Lu\|_{L_w^1(B_1(0))} \leq C\left(w(B_1(0)) \sum_{0 \leq |\alpha| \leq k} \|\mu_\alpha\|_{L_{w,1}^{1,N-(m-\alpha)}(\mathbf{R}^N)} + C_k\right) \|u\|_{W_w^{m,1}(B_1(0))},$$

where $C_k = 1$, if $k \leq m-2$, and $C_k = 0$, if $k = m-1$. The constant C depends only on k, m, N, and the A_1 constant of w. The projection L is uniquely determined by the requirement that

$$\int_{\overline{B_1(0)}} D^\alpha u \, d\mu_\alpha = \int_{\overline{B_1(0)}} D^\alpha(Lu) \, d\mu_\alpha, \quad |\alpha| \leq m-1.$$

Lemma 4.3.9. *Let k and m be integers, $0 \leq k \leq m - 1$, and let $w \in A_1$. Let $u \in W_w^{m,1}(B_1(0))$, and suppose that $D^\alpha u|_E = 0$ for $0 \leq |\alpha| \leq k$, where $E \subset B_1(0)$ is a Suslin set with $\mathcal{H}_{w,1}^{N-(m-k)}(E) > 0$. Then there exists a polynomial $P \in \mathcal{P}_{m-1}$, $P(x) = \sum_{|\beta| \leq m-1} a_\beta x^\beta$, such that*

$$\|u - Lu\|_{W_w^{m,1}(B_1(0))} \leq C \frac{w(B_1(0))}{\mathcal{H}_{w,1}^{N-(m-k)}(E)} \|\nabla^m u\|_{L_w^1(B_1(0))}$$

and

$$|a_\beta| \leq C w(B_1(0)) \|\nabla^{k+1} u\|_{L_w^1(B_1(0))}, \quad 0 \leq |\beta| \leq m - 1.$$

In the case $k = m - 1$, P is identically 0. The constant C depends only on k, m, N, and the A_1 constant of w.

Corollary 4.3.10. *Let m be an integer and $w \in A_1$. Then there is a constant C, depending only on m, N, and the A_1 constant of w, such that if $u \in W_w^{m,1}(B_r(a))$ and $D^\alpha u|_E = 0$ for $0 \leq |\alpha| \leq m - 1$, where $E \subset B_r(a)$ is a Suslin set with $\mathcal{H}_{w,r}^{N-1}(E) > 0$, then*

$$\int_{B_r(a)} |u| w \, dx \leq C \frac{w(B_r(a))}{r \mathcal{H}_{w,r}^{N-1}(E)} r^m \int_{B_r(a)} |\nabla^m u| w \, dx. \tag{4.3.11}$$

Corollary 4.3.11. *Let k and m be integers, $0 \leq k \leq m - 1$, and let $w \in A_1$. Then there is a constant C, depending only on k, m, N, and the A_1 constant of w, such that if $u \in W_w^{m,1}(B_r(a))$ and $D^\alpha u|_E = 0$ for $0 \leq |\alpha| \leq k$, where $E \subset B_r(a)$ is a Suslin set with $\mathcal{H}_{w,r}^{N-(m-k)}(E) > 0$, then*

$$\int_{B_r(a)} |u| w \, dx \leq C r^{k+1} \int_{B_r(a)} |\nabla^{k+1} u| w \, dx$$

$$+ C \frac{r^{-(m-k)} w(B_r(a))}{\mathcal{H}_{w,r}^{N-(m-k)}(E)} r^m \int_{B_r(a)} |\nabla^m u| w \, dx.$$

4.4. Spectral synthesis

The starting point for this book was the following theorem by L. I. Hedberg.

Theorem 4.4.1. *Let $u \in W^{m,p}(\mathbf{R}^N)$, where $1 < p < \infty$. Let $\Omega \subset \mathbf{R}^N$ be open, and suppose that $D^\alpha u = 0$ $B_{m-|\alpha|,p}$-quasieverywhere on Ω^c for $|\alpha| \leq m - 1$. Then u belongs to the closure of $C_0^\infty(\Omega)$ in $W^{m,p}(\mathbf{R}^N)$.*

Let us briefly discuss the history of this theorem and some later developments. A detailed account can be found in Adams–Hedberg [7, pp. 288–289].

The theorem goes back at least to S. L. Sobolev [96] (see also Sobolev's book [98]), who, in connection with a uniqueness theorem for the polyharmonic equation, considered the case when Ω is bounded by smooth manifolds. For the case when Ω satisfies the segment property, see Folland [41, p. 278].

With $m = 1$ and $p = 2$, the theorem is due to A. Beurling [15] and J. Deny [31], and with $m = 1$ and $1 < p < \infty$ to V. P. Havin [52] and T. Bagby [11]. For $m \geq 2$, partial results were obtained by V. I. Burenkov [16], J. C. Polking [93], and Hedberg [55]. With the additional assumption that $p > 2 - 1/N$, the theorem was then proved by Hedberg in [56], [57]. The reason for this restriction was the fact the Kellogg property for $B_{\alpha,p}$-thinness, at that time, was only known to be true for $p > 2 - \alpha/N$. Later, Th. H. Wolff [58] established the Kellogg property for general p, $1 < p < \infty$, which proved Theorem 4.4.1.

Yu. V. Netrusov has given an entirely different proof of Theorem 4.4.1 which is valid for a much more general class of spaces, possessing an unconditional base, including Besov and Triebel–Lizorkin spaces [89]; see also Chapter 10 in Adams–Hedberg [7].

For $p = 1$, a corresponding theorem was proved in the special case when Ω^c has $\mathcal{H}_\infty^{N-m}$-capacity 0 by A. Carlsson [21] and in full generality by Netrusov [90].

Hedberg's result has been extended to A_p weighted Sobolev spaces with $1 < p < \infty$ by N. O. Belova [12], [14].

Below, we shall prove two weighted versions of Hedberg's theorem, one for $1 < p < \infty$ and one for $p = 1$. In both cases it will be assumed that Ω^c satisfies an extra "thickness" condition. We shall also show that these conditions are fulfilled, e.g., if Ω is a bounded Lipschitz domain.[†]

4.4.1. The case $1 < p < \infty$

In the non-weighted case, the following theorem is due to L. I. Hedberg; see Adams–Hedberg [7, pp. 251–253].

Theorem 4.4.2. *Let Ω be an open subset of $\mathbf{R}^N$, and suppose that $K = \Omega^c$ satisfies the following uniform thickness condition: there exists a sequence $\{a_n\}_{n=1}^\infty$ of positive numbers, for which*

$$\sum_{n=1}^\infty a_n^{p'} = \infty \tag{4.4.1}$$

and such that

$$\frac{R_{1,p;2^{-n}}^w (K \cap \overline{B_{2^{-n}}(x)})}{2^{np} w(B_{2^{-n}}(x))} \geq a_n^p, \quad n = 1, 2, \ldots, \tag{4.4.2}$$

for every $x \in K$. Let $u \in W_w^{m,p}(\mathbf{R}^N)$, and suppose that $D^\alpha u|_K = 0$, $|\alpha| \leq m-1$. Then u belongs to the closure of $C_0^\infty(\Omega)$ in $W_w^{m,p}(\mathbf{R}^N)$.

[†]The proof by Hedberg in [54] for the case $m = 1$ also works *mutatis mutandis* for weighted Sobolev spaces. We shall not, however, go into this here.

In the proof of this theorem, we shall use a simple lemma.

Lemma 4.4.3. *Let $0 < \alpha < N$ and $1 < p < \infty$, and let w be an A_p weight. Suppose that $E \subset \mathbf{R}^N$ satisfies $R^w_{\alpha,p;1}(E) = 0$. Then $|E| = 0$.*

Proof. We may assume that E is a subset of an open ball B with radius 1. Let G be open such that $E \subset G \subset 2B$, and let $f \geq 0$ satisfy $\mathcal{I}_{\alpha,1}f \geq 1$ on E. In the same way as in the proof of the inequality (2.1.11), it follows that

$$w(G) \leq \int_{2B} (\mathcal{I}_{\alpha,1}f)^p w\, dx \leq C \int_{2B} f^p w\, dx.$$

But since the last integral can be made arbitrarily small, we get $w(G) = 0$, and thus $|G| = 0$. This fact implies that the exterior measure of E is 0, which proves the lemma. $\square$

Proof of Theorem 4.4.2. We begin with some preliminary reductions. First of all, we can assume that K is compact. For if $\chi_R \in C_0^\infty(B_{2R}(0))$ satisfies $\chi_R = 1$ on $\overline{B_R(0)}$, then by the Leibniz rule, $D^\alpha(\chi_R u) = 0$ on the compact set $K \cap \overline{B_R(0)}$ for $|\alpha| \leq m - 1$. It is also easy to see that $\chi_R u \to u$ in $W_w^{m,p}(\mathbf{R}^N)$, as $R \to \infty$. Now let V be an arbitrary neighbourhood of K. We shall construct a function $\omega \in C_0^\infty(V)$ such that $\omega = 1$ in a smaller neighbourhood of K and

$$\|\omega u\|_{W_w^{m,p}(\mathbf{R}^N)} \leq C \int_V |\nabla^m u|^p w\, dx. \qquad (4.4.3)$$

Note that the last integral can be made arbitrarily small just by choosing V close enough to K. Indeed, by the assumption and Lemma 4.4.3, $D^\alpha u = 0$ a.e. on K for every α with $|\alpha| = m - 1$. But if $|\alpha| = m - 1$, then $D^\alpha u \in W_w^{1,p}(\mathbf{R}^N)$, so it follows from the truncation theorem, Theorem 2.1.7, that $D^\beta u = 0$ on K for every β with $|\beta| = m$, and hence

$$\int_V |\nabla^m u|^p w\, dx = \int_{V \setminus K} |\nabla^m u|^p w\, dx.$$

If we now define $\eta = 1 - \omega$, then $\eta \in C_0^\infty(\mathbf{R}^N)$, and

$$\|u - \eta u\|_{W_w^{m,p}(\mathbf{R}^N)} \leq C \int_V |\nabla^m u|^p w\, dx.$$

The function ηu can in turn, according to Corollary 2.1.5, be approximated arbitrarily well by functions belonging to $C_0^\infty(\Omega)$.

We now proceed to construct the multiplier ω. For $n = 1, 2, \ldots$, we let

$$G_n = \{x \in \mathbf{R}^N \,;\, \mathrm{dist}(x, K) < 2^{-(n+1)}\}.$$

Let $\omega_n \in C_0^\infty(G_n)$ satisfy $\omega_n = 1$ on G_{n+1}, $0 \leq \omega_n \leq 1$, and $|D^\alpha \omega_n| \leq C 2^{|\alpha| n}$ for $|\alpha| \leq m$. For a fixed M, we then choose P so large that

$$1 \leq \sum_{n=M}^P a_n^{p'} \leq C.$$

If we now define the function ω by

$$\omega = \frac{\sum_{n=M}^{P} a_n^{p'} \omega_n}{\sum_{n=M}^{P} a_n^{p'}},$$

then $\omega \in C_0^\infty(G_M)$, $0 \leq \omega \leq 1$, $\omega = 1$ on G_{P+1}, and $|D^\alpha \omega| \leq a_n^{p'} |D^\alpha \omega_n| \leq C 2^{|\alpha| n} a_n^{p'}$ on $G_n \setminus G_{n+1}$ for $M \leq n \leq P$.

For notational simplicity, we shall temporarily denote the ball $B_{2^{-n}}(x)$ by $B_n(x)$. For every $x_0 \in G_n \setminus G_{n+1}$, there exists a point $\bar{x} \in K$ such that $B_{n+2}(x_0) \subset B_n(\bar{x}) \subset B_{n-1}(x_0)$. First notice that

$$\int_{\mathbf{R}^N} \omega^p |\nabla^m u|^p w \, dx \leq \int_{G_M} |\nabla^m u|^p w \, dx \leq \int_V |\nabla^m u|^p w \, dx,$$

if M is large enough. Let k and l be integers, $0 \leq l \leq k \leq m-1$. Applying the Poincaré inequality (4.3.9), we then find

$$\int_{B_{n+2}(x_0)} |\nabla^{k-l} u|^p w \, dx \leq \int_{B_n(\bar{x})} |\nabla^{k-l} u|^p w \, dx$$

$$\leq C \frac{2^{np} w(B_n(\bar{x}))}{R_{1,p;2^{-n}}^w(K \cap \overline{B_n(\bar{x})})} 2^{-(m-(k-l))np} \int_{B_n(\bar{x})} |\nabla^m u|^p w \, dx$$

$$\leq C a_n^{-p} 2^{-lnp} \int_{B_{n-1}(x_0)} |\nabla^m u|^p w \, dx. \qquad (4.4.4)$$

We now cover $G_n \setminus G_{n+1}$ with balls $\{B_{n+2}(x_i)\}$, where $x_i \in G_n \setminus G_{n+1}$, such that $\{B_{n-1}(x_i)\}$ covers $G_n \setminus G_{n+1}$ with finite multiplicity only depending on N. The inequality (4.4.4) then yields

$$\int_{\mathbf{R}^N} |\nabla^l \omega|^p |\nabla^{k-l} u|^p w \, dx = \sum_{n=M}^{P} \int_{G_n \setminus G_{n+1}} |\nabla^l \omega|^p |\nabla^{k-l} u|^p w \, dx$$

$$\leq C \sum_{n=M}^{P} 2^{lnp} a_n^{pp'} \int_{B_{n+2}(x_i)} |\nabla^{k-l} u|^p w \, dx$$

$$\leq C \sum_{n=M}^{P} a_n^{p'} \int_{B_{n-1}(x_i)} |\nabla^m u|^p w \, dx$$

$$\leq C \int_{G_{M-3}} |\nabla^m u|^p w \, dx.$$

Finally, note that

$$\int_{G_{M-3}} |\nabla^m u|^p w \, dx \leq \int_V |\nabla^m u|^p w \, dx$$

if M is large enough. This proves (4.4.3) and the theorem. $\square$

We now give a sufficient condition for the set $K = \Omega^c$ to be uniformly thick.

Proposition 4.4.4. *Let* $0 < \alpha < N$ *and* $1 < p < \infty$, *and let* w *be an* A_p *weight. Let* E *be a measurable subset of* $\overline{B_r(a)}$ *for some ball* $B_r(a)$, *and suppose that* $|E| \geq \rho|B_r(a)|$. *Then*

$$R^w_{\alpha,p;r}(E) \geq C\rho^p R^w_{\alpha,p;r}(B_r(a)),$$

where the constant C *only depends on* α, N, p, *and the* A_p *constant of* w.

Proof. Suppose that $f \geq 0$ and that $\mathcal{I}_{\alpha,r}f \geq 1$ on E. Then, as in the proof of Lemma 3.3.12,

$$w(E) \leq Cr^{\alpha p} \int_{\mathbf{R}^N} f^p w \, dx.$$

By the strong doubling property of w and Proposition 3.3.12, we also have

$$w(E) \geq C\rho^p w(B_r(a)) \geq C\rho^p r^{\alpha p} R^w_{\alpha,p;r}(B_r(a)).$$

It follows that

$$\int_{\mathbf{R}^N} f^p w \, dx \geq C\rho^p R^w_{\alpha,p;r}(B_r(a)),$$

which proves the proposition. $\square$

Suppose that there exists a constant ρ, $0 < \rho \leq 1$, such that

$$|K \cap \overline{B_r(x)}| \geq \rho|B_r(x)|$$

for every $x \in K$ and every $r \leq 1$. Proposition 3.3.12 and Proposition 4.4.4 then imply

$$\frac{R^w_{1,p;2^{-n}}(K \cap \overline{B_{2^{-n}}(x)})}{2^{np}w(B_{2^{-n}}(x))} \geq C\rho^p \frac{R^w_{1,p;2^{-n}}(B_{2^{-n}}(x))}{2^{np}w(B_{2^{-n}}(x))} \geq C$$

for every $x \in K$ and $n = 0, 1, \ldots$. Thus, K is uniformly thick.

4.4.2. The case $p = 1$

Theorem 4.4.5. *Let* m *be an integer,* $1 \leq m < N$, *and let* $w \in A_1$. *Let* Ω *be an open subset of* $\mathbf{R}^N$, *and suppose that* $K = \Omega^c$ *satisfies the following capacitary cone condition: there exist two positive numbers* a *and* R *such that*

$$\frac{r\mathcal{H}^{N-1}_{w,r}(K \cap \overline{B_r(x)})}{w(B_r(x))} \geq a \tag{4.4.5}$$

for every $x \in K$ *and* $0 < r \leq R$. *Let* $u \in W^{m,1}_w(\mathbf{R}^N)$, *and suppose that* $D^\alpha u|_K = 0$ *for* $0 \leq |\alpha| \leq m - 1$. *Then* u *belongs to the closure of* $C^\infty_0(\Omega)$ *in* $W^{m,1}_w(\mathbf{R}^N)$.

Proof. As in the proof of Theorem 4.4.2, we can assume that K is compact. Given an arbitrary neighbourhood V of K, it also suffices to construct a function $\omega \in C_0^\infty(V)$ such that $\omega = 1$ in a smaller neighbourhood of K and

$$\|\omega u\|_{W_w^{m,1}(\mathbf{R}^N)} \leq C \int_V |\nabla^m u| w \, dx. \tag{4.4.6}$$

For $n = 1, 2, \ldots$, let $G_n = \{x \in \mathbf{R}^N \; ; \; \mathrm{dist}(x, K) < 2^{-(n+1)}\}$. Let $\omega \in C_0^\infty(V)$ satisfy $\omega = 1$ on G_{n+1} and $|D^\alpha \omega| \leq C2^{|\alpha|n}$ for $|\alpha| \leq m$. For $x_0 \in G_n \setminus G_{n+1}$, there exists a point $\bar{x} \in K$ such that $B_{n+2}(x_0) \subset B_n(\bar{x}) \subset B_{n-1}(x_0)$ (here, we use the same notation as before). Let k and l be integers, $0 \leq l \leq k \leq m-1$, and suppose that n is so large that $2^{-n} \leq R$. The Poincaré inequality (4.3.11) implies that

$$\int_{B_{n+2}(x_0)} |\nabla^{k-l} u| w \, dx \leq \int_{B_n(\bar{x})} |\nabla^{k-l} u| w \, dx$$

$$\leq C \frac{2^n w(B_n(\bar{x}))}{\mathcal{H}_{w,2^{-n}}^{N-1}(K \cap B_n(\bar{x}))} 2^{-(m-(k-l))n} \int_{B_n(\bar{x})} |\nabla^m u| w \, dx$$

$$\leq \frac{C}{a} 2^{-ln} \int_{B_{n-1}(x_0)} |\nabla^m u| w \, dx. \tag{4.4.7}$$

We cover $G_n \setminus G_{n+1}$ with balls $\{B_{n+2}(x_i)\}$, where $x_i \in G_n \setminus G_{n+1}$, such that $\{B_{n-1}(x_i)\}$ covers $G_n \setminus G_{n+1}$ with finite multiplicity. It follows from (4.4.7) that

$$\int_{\mathbf{R}^N} |\nabla^l \omega| |\nabla^{k-l} u| w \, dx \leq C2^{ln} \sum_i \int_{B_{n+2}(x_i)} |\nabla^{k-l} u| w \, dx$$

$$\leq C \sum_i \int_{B_{n-1}(x_i)} |\nabla^m u| w \, dx$$

$$\leq C \int_{G_{n-3}} |\nabla^m u| w \, dx$$

$$\leq C \int_V |\nabla^m u| w \, dx,$$

if n is large enough. $\square$

The proof of Theorem 3.5.5 can be used *verbatim* to produce the following result.

Lemma 4.4.6. *Let m be an integer, $1 \leq m < N$, and let $w \in A_1$. Suppose that K is a compact subset of $\overline{B_r(a)}$. Then the capacity $\mathcal{H}_{w,r}^{N-m}(K)$ is comparable to the varitional capacity*

$$\inf\{\|\varphi\|_{W_w^{m,1}(\mathbf{R}^N)} \; ; \; \varphi \in C_0^\infty(B_{4r}(a)), \, 0 \leq \varphi \leq 1, \, and$$
$$\varphi = 1 \text{ in a neighbourhood of } E\}.$$

Proposition 4.4.7. *Let* $0 < \alpha < N$, *and let* w *be an* A_1 *weight. Let* E *be a measurable subset of* $\overline{B_r(a)}$ *for some ball* $B_r(a)$, *and suppose that* $|E| \geq \rho|B_r(a)|$. *Then*

$$\mathcal{H}_{w,r}^{N-\alpha}(E) \geq C\rho^{(N-m)/N}\mathcal{H}_{w,r}^{N-\alpha}(B_r(a)),$$

where the constant C *only depends on* α, N, *and the* A_1 *constant of* w.

Proof. Let $\varphi \in C_0^\infty(B_{4r}(a))$ be arbitrary such that $0 \leq \varphi \leq 1$ and $\varphi = 1$ in a neighbourhood of E. The strong doubling property of w and the Sobolev type inequality in Theorem 2.5.2 imply that

$$
\begin{aligned}
w(B_r(a))^{(N-m)/N} &\leq C\left(\frac{|B_r(a)|}{|E|}\right)^{(N-m)/N} w(E)^{(N-m)/N} \\
&\leq C\rho^{-(N-m)/N}w(E)^{(N-m)/N} \\
&\leq C\rho^{-(N-m)/N}\left(\int_{B_{4r}(a)} |\varphi|^{N/(N-m)}w\,dx\right)^{(N-m)/N} \\
&\leq C\rho^{-(N-m)/N}\left(\frac{|B_{4r}(a)|}{w(B_{4r}(a))}\right)^{m/N}\int_{B_{4r}(a)} |\nabla^m\varphi|w\,dx \\
&\leq C\rho^{-(N-m)/N}r^m w(B_r(a))^{-m/N}\|\varphi\|_{W_w^{m,1}(\mathbf{R}^N)},
\end{aligned}
$$

whence

$$\|\varphi\|_{W_w^{m,1}(\mathbf{R}^N)} \geq C\rho^{(N-m)/N}\frac{w(B_r(a))}{r^m} \geq C\rho^{(N-m)/N}\mathcal{H}_{w,r}^{N-\alpha}(B_r(a)).$$

The assertion now follows from Lemma 4.4.6. $\square$

In the same way as in the previous section, this proposition and Theorem 3.4.6 imply that K satisfies the assumption (4.4.5), if

$$|K \cap \overline{B_r(x)}| \geq \rho|B_r(x)|$$

for every $x \in K$ and $0 < r \leq 1$.

References

1. ADAMS, D. R., Traces of potentials arising from translation invariant operators, *Ann. Scuola Norm. Sup. Pisa Cl. Sci.* **25** (1971), 203–217.

2. ADAMS, D. R., A trace inequality for generalized potentials, *Studia Math.* **48** (1973), 99–105.

3. ADAMS, D. R., Traces of potentials. II, *Indiana Univ. Math. J.* **22** (1973), 907–918.

4. ADAMS, D. R., A note on the Choquet integral with respect to Hausdorff capacity, in *Function Spaces and Applications, Proceedings, Lund 1986* (M. Cwikel et al., eds.), *Lecture Notes in Math.* **1302**, pp. 115–124, Springer-Verlag, Berlin–Heidelberg, 1986.

5. ADAMS, D. R., Weighted nonlinear potential theory, *Trans. Amer. Math. Soc.* **297** (1986), 73–94.

6. ADAMS, D. R., Weighted capacity and the Choquet integral, *Proc. Amer. Math. Soc.* **102** (1988), 879–887.

7. ADAMS, D. R. and HEDBERG, L. I., *Function Spaces and Potential Theory*, Springer-Verlag, Heidelberg–New York, 1996.

8. ADAMS, D. R. and MEYERS, N. G., Thinness and Wiener criteria for non-linear potentials, *Indiana Univ. Math. J.* **22** (1972), 169–197.

9. ADAMS, R. A., *Sobolev Spaces*, Academic Press, New York, 1975.

10. AIKAWA, H., On weighted Beppo Levi functions—Integral representations and behaviour at infinity, *Analysis* **9** (1989), 323–346.

11. BAGBY, T., Quasi topologies and rational approximation, *J. Funct. Anal.* **10** (1972), 259–268.

12. BELOVA, N. O., Spectral synthesis in weighted Sobolev spaces, *Mat. Zametki* **56** (1994), 136–139 (Russian). English transl.: *Math. Notes* **56** (1994), 856–858.

13. BELOVA, N. O., The weighted Poincaré inequality, in *Analysis and discrete mathematics* (M. M. Lavrentiev, ed.), pp. 20–29, Novosibirsk. Gos. Univ., Novosibirsk, 1995 (Russian).

14. BELOVA, N. O., Spectral synthesis in weighted Sobolev spaces, *Trudy Instituta Matematiki* **31** (1996), 3–39 (Russian). English transl.: *Siberian Adv. Math.* **8** (1998), 1–48.

15. BEURLING, A., Sur les spectres des fontions, in *Analyse harmonique*, Colloque, Nancy 1947, 9–29, Centre National de la Recherche Scientifique, Paris, 1949. Reprinted in *The Collected Works of Arne Beurling, Vol. 2, Harmonic Analysis*, 125–145, Birkhäuser, Boston, Mass., 1989.

16. BURENKOV, V. I., On the approximation of functions in the space $W_p^r(\Omega)$ by functions with compact support for an arbitrary open set Ω, *Trudy. Mat. Inst. Steklov.*, **131** (1974) 51–63 (Russian). English transl.: *Proc. Steklov Inst. Math.* **131** (1974) 51–63.

17. BURKHOLDER, D. L. and GUNDY, R. F., Extrapolation and interpolation of quasilinear operators on martingales, *Acta Math.* **124** (1970), 249–304.

18. CALDERÓN, A. P., Lebesgue spaces of differentiable functions and distributions, in *Partial Differential Equations* (C. B. Morrey Jr, ed.), *Proc. Symp. Pure Math.* 4, pp. 33–49, Amer. Math. Soc., Providence, R. I., 1961.

19. CARLESON, L., *Selected Problems on Exceptional Sets*, Van Nostrand, Princeton, N. J., 1967.

20. CARLSSON, A., Inequalities of Poincaré–Wirtinger type, *Licentiate Thesis* no. 232, Dept. of Math., Linköping Univ., Linköping, 1990.

21. CARLSSON, A., A note on spectral synthesis, *LiTH-MAT-R-93-03*, Dept. of Math., Linköping Univ., Linköping, 1993.

22. CARLSSON, A. and MAZ'YA, V. G., On approximation in weighted Sobolev spaces and self-adjointness, *Math. Scand.* **74** (1994), 111–124.

23. CHOQUET, G., Theory of capacities, *Ann. Inst. Fourier (Grenoble)* **5** (1955), 131–255.

24. CHUA, S. K., Extension Theorems on Weighted Sobolev Spaces, Ph.D. Thesis, Rutgers University, Rutgers, N. J., 1990.

25. CHUA, S. K., Extension theorems on weighted Sobolev spaces, *Indiana Univ. Math. J.* **41** (1992), 1027–1076.

26. CHUA, S. K., Some remarks on extension theorems for weighted Sobolev spaces, *Illinois J. Math.* **38** (1994), 95–126.

27. CLARKSON, J. A., Uniformly convex spaces, *Trans. Amer. Math. Soc.* **40** (1936), 396–414.

28. COIFMAN, R. R. and FEFFERMAN, C., Weighted norm inequalities for maximal functions and singular integrals, *Studia Math.* **51** (1974), 241–250.

29. COIFMAN, R. R. and ROCHBERG, R., Another characterization of BMO, *Proc. Amer. Math. Soc.* **79** (1980), 249–254.

30. DAVID, G. and SEMMES, S., Strong A_∞ weights, Sobolev inequalities and quasiconformal mappings, in *Analysis and Partial Differential Equations* (C. Sadosky, ed.), pp. 101–111, Marcel Dekker, New York, 1990.

31. DENY, J., Les potentiels d'énergie finie, *Acta Math.* **82** (1950), 107–183.

32. DENY, J. and LIONS, J. L., Les espaces du type de Beppo Levi, *Ann. Inst. Fourier (Grenoble)* **5** (1953–54), 305–370.

33. EVANS, G. C., Applications of Poincaré's sweeping-out process, *Proc. Nat. Acad. USA* **37** (1935), 226–253.

34. EVANS, L. C. and GARIEPY, R. F., *Measure Theory and Fine Properties of Functions*, CRC Press, Boca Raton, Fla., 1992.

35. FABES, E. B., JERISON, D. S., and KENIG, C. E., The Wiener test for degenerate elliptic equations, *Ann. Inst. Fourier (Grenoble)* **32** (1982), 151–182.

36. FABES, E. B., KENIG, C. E., and SERAPIONI, R. P., The local regularity of solutions of degenerate elliptic equations, *Comm. Partial Differential Equations* **7** (1982), 77–116.

37. FABES, E. B., KENIG, C. E., and JERISON, D. S. Boundary behaviour of solutions to degenerate elliptic equations, in *Conference on Harmonic Analysis in Honor of Antoni Zygmund*, Proc., Chicago 1981 (W. Beckner, R. Fefferman, P. W. Jones, eds.), 577–589, Wadsworth, Belmont, Calif., 1983.

38. FEDERER, H., *Geometric Measure Theory*, Springer-Verlag, Heidelberg-New York, 1969.

39. FEDERER, H. and FLEMING, W. H., Normal and integral currents, *Ann. of Math.* **72** (1960), 458–520.

40. FERNSTRÖM, C., On the instability of Hausdorff content, *Math. Scand.* **60** (1987), 19–30.

41. FOLLAND, G. B., *Introduction to Partial Differential Equations*, Princeton Univ. Press, Princeton, N. J., 1976.

42. FROSTMAN, O., Potentiel d'équilibre et capacité des ensembles avec quelques applications à la théorie des fonctions, *Medd. Lunds Univ. Mat. Sem.* **3** (1935), 1–118.

43. FUGLEDE, B., On the theory of potentials in locally compact spaces, *Acta Math.* **103** (1960), 139–215.

44. GAGLIARDO, E., Proprieta di alcune classi di funzioni in piu variabili, *Ricerche Mat.* **7** (1958), 102–137.

45. GARCÍA-CUERVA, J. and RUBIO DE FRANCIA, J. L., *Weighted Norm Inequalities and Related Topics*, North-Holland, Amsterdam, 1985.

46. GIAQUINTA, M., *Multiple Integrals in the Calculus of Variations and Non-linear Elliptic Systems*, Princeton Univ. Press, Princeton, N. J., 1983.

47. GILBARG, D. and TRUDINGER, N. S., *Elliptic Partial Differential Equations of Second Order*, Springer-Verlag, New York, Second edition, 1983.

48. GUNDY, R. F. and WHEEDEN, R. L., Weighted inequalities for the non-tangential maximal function, Lusin area integral, and Walsh–Paley series, *Studia Math.* **49** (1974), 107–124.

49. GUSTIN, W., Boxing inequalities, *J. Math. Mech.* **9** (1960), 229–239.

50. HARDY, G. H. and LITTLEWOOD, J. E., Some properties of fractional integrals, *Math. Z.* **27** (1928), 565–606.

51. HARVEY, R. and POLKING, J. C., Removable singularities of solutions of linear partial differential equations, *Acta Math.* **125** (1970), 39–56.

52. HAVIN, V. P., Approximation in the mean by analytic functions, *Dokl. Akad. Nauk. SSSR* **178** (1968), 1025–1028 (Russian). English transl.: *Soviet Math. Dokl.* **9** (1968), 245–248.

53. HAVIN, V. P. and MAZ'YA, V. G., Nonlinear potential theory, *Uspekhi Mat. Nauk* 27:6 (1972), 67–138 (Russian). English transl.: *Russian Math. Surveys* **27** (1972), 71–148.

54. HEDBERG, L. I., Non-linear potentials and approximation in the mean by analytic functions, *Math. Z.* **129** (1972), 299–319.

55. HEDBERG, L. I., Approximation in the mean by solutions of elliptic equations, *Duke Math. J.* **40** (1973), 9–16.

56. HEDBERG, L. I., Two approximation problems in function spaces, *Ark. mat.* **16** (1978), 51–81.

57. HEDBERG, L. I., Spectral synthesis in Sobolev spaces, and uniqueness of solutions of the Dirichlet problem, *Acta Math.* **147** (1981), 237–264.

58. HEDBERG, L. I. and WOLFF, TH. H., Thin sets in nonlinear potential theory, *Ann. Inst. Fourier (Grenoble)* **33** (1983), 161–187.

59. HEINONEN, J., KILPELÄINEN, T., and MARTIO O., *Nonlinear Potential Theory of Degenerate Elliptic Equations,* Oxford Univ. Press, Oxford, 1993.

60. HEWITT, E. and STROMBERG, K., *Real and Abstract Analysis*, Springer-Verlag, Berlin–Heidelberg, 1965.

61. HUNT, R., MUCKENHOUPT, B., and WHEEDEN, R. L., Weighted norm inequalities for the conjugate function and Hilbert transform, *Trans. Amer. Math. Soc.* **176** (1973), 227–251.

62. JONES, P. W., Quasiconformal mappings and extendability of functions in Sobolev spaces, *Acta Math.* **147** (1981), 71–88.

63. KALTON, N. J. and VERBITSKY, I. E., Nonlinear equations and weighted norm inequalities, *Trans. Amer. Math. Soc.* **351** (1999), 3441–3497.

64. KAMETANI, S., On some properties of Hausdorff's measure and the concept of capacity in generalized potentials, *Proc. Imp. Acad. Tokyo* **18** (1942), 617–625.

65. KELLOGG, O. D., Unicité des fonctions harmoniques, *C. R. Acad. Sci. Paris* **187** (1928), 526–527.

66. KELLOGG, O. D., *Foundations of potential theory*, Springer-Verlag, Berlin, 1929.

67. KILPELÄINEN, T., Weighted Sobolev spaces and capacity, *Ann. Acad. Sci. Fenn. Ser. A I Math.* **19** (1994), 95–113.

68. KILPELÄINEN, T. and MALÝ, J., The Wiener test and potential estimates for quasilinear elliptic equations, *Acta Math.* **172** (1994), 137–161.

69. LANDKOF, N. S., *Foundations of Modern Potential Theory*, Nauka, Moscow, 1966 (Russian). English transl.: Springer-Verlag, Berlin–Heidelberg, 1972.

70. MAZ'YA, V. G., Classes of domains and imbedding theorems for function spaces, *Dokl. Akad. Nauk SSSR* **133** (1960), 527–530 (Russian). English transl.: *Soviet Math. Dokl.* **1** (1960), 882–885.

71. MAZ'YA, V. G., The Dirichlet problem for elliptic equations of arbitrary order in unbounded regions, *Dokl. Akad. Nauk SSSR* **150** (1963), 1221–1224 (Russian). English transl.: *Soviet Math.* **4** (1963), 860–863.

72. MAZ'YA, V. G., The continuity at a boundary point of the solutions of quasi-linear elliptic equations, *Vestnik Leningrad. Univ.* **25** (1970), 42–55 (Russian. English summary). Correction: *ibid.* **27** (1972), 160. English transl.: *Vestnik Leningrad. Univ.* **3** (1976), 225–242.

73. MAZ'YA, V. G., On certain integral inequalities for functions of many variables, *Problemy Matematicheskogo Analiza, Leningrad. Univ.* **3** (1972), 33–68 (Russian). English transl.: *J. Soviet Math.* **1** (1973), 205–234.

74. MAZ'YA, V. G., On (p, l)-capacity, imbedding theorems and the spectrum of a selfadjoint elliptic operator, *Izv. Akad. Nauk SSSR, Ser. Mat.* **37** (1973), 356–385 (Russian). English transl.: *Math. USSR-Izv.* **7** (1973), 357–387.

75. MAZ'YA, V. G., Strong capacity-estimates for "fractional" norms, *Zap. Nauchn. Semin. Leningrad. Otdel. Mat. Inst. Steklov. (LOMI)* **70** (1977), 161–168 (Russian). English transl.: *J. Soviet Math.* **23** (1983), 1997–2003.

76. MAZ'YA V. G., *Sobolev Spaces*, Springer-Verlag, Berlin–Heidelberg, 1985.

77. MEYERS, N. G., A theory of capacities for potentials of functions in Lebesgue classes, *Math. Scand.* **26** (1970), 255–292.

78. MEYERS, N. G., Continuity properties of potentials, *Duke Math. J.* **42** (1975), 157–166.

79. MEYERS, N. G., Integral inequalities of Poincaré and Wirtinger type, *Arch. Rational Mech. Anal.* **68** (1978), 113–120.

80. MEYERS, N. G. and SERRIN, J., $H = W$, *Proc. Nat. Acad. Sci. U.S.A.* **51** (1964), 1055–1056.

81. MEYERS, N. G. and ZIEMER, W. P., Integral inequalities of Poincaré and Wirtinger type for BV functions, *Amer. J. Math.* **99** (1977), 1345–1360.

82. MILLER, N., Weighted Sobolev spaces and pseudodifferential operators with smooth symbols, *Trans. Amer. Math. Soc.* **269** (1982), 91–109.

83. MORREY JR, C. B., On the solution of quasi-linear partial differential equations, *Trans. Amer. Math. Soc.* **43** (1938), 126–166.

84. MORSE, A. P., The behaviour of a function on its critical set, *Ann. of Math.* **40** (1939), 62–70.

85. MUCKENHOUPT, B., Weighted norm inequalities for the Hardy maximal function, *Trans. Amer. Math. Soc.* **165** (1972), 207–226.

86. MUCKENHOUPT, B., The equivalence of two conditions for weight functions, *Studia Math.* **49** (1974), 101–106.

87. MUCKENHOUPT, B. and WHEEDEN, R. L., Weighted norm inequalities for fractional integrals, *Trans. Amer. Math. Soc.* **192** (1974), 261–274.

88. MUCKENHOUPT, B. and WHEEDEN, R. L., On the dual of weighted H^1 on the half-space, *Studia Math.* **53** (1978), 57–79.

89. NETRUSOV, YU. V., Spectral synthesis in spaces of smooth functions, *Ross. Akad. Nauk Dokl.* **325** (1992), 923–925 (Russian). English transl.: *Russian Acad. Sci. Dokl. Math.* **46** (1993), 135–137.

90. NETRUSOV, YU. V., Spectral synthesis in the Sobolev space associated with integral metric, *Zap. Nauchn. Semin. S.-Petersburg. Otdel. Mat. Inst. Steklov. (POMI)* **217** (1994), 217–234 (Russian). English tranl.: *J. Math. Sci.*, to appear.

91. NIEMINEN, E., Hausdorff Measures, Capacities, and Sobolev Spaces with Weights, *Ann. Acad. Sci. Fenn. Ser. A I Math. Dissertationes* **81**, Helsinki, 1991.

92. OSSERMAN, R., The isoperimetric inequality, *Bull. Amer. Math. Soc.* **84** (1978), 1182–1238.

93. POLKING, J. C., Approximation in L^p by solutions of elliptic partial differential equations, *Amer. J. Math.* **94** (1972), 1231–1244.

94. RESHETNYAK, YU. G., The concept of capacity in the theory of functions with generalized derivatives, *Sibirsk. Mat. Zh.* **10** (1969), 1109–1138 (Russian). English transl.: *Siberian Math. J.* **10** (1969), 818–842.

95. RUDIN, W., *Real and Complex Analysis*, McGraw-Hill, New York, Second edition, 1974.

96. SOBOLEV, S. L., On a boundary value problem for polyharmonic equations, *Mat. Sb. (N. S.)* **2** (**44**) (1937), 465–499 (Russian). English transl.: *Amer. Math. Soc. Translations* (2) **33** (1963), 1–40.

97. SOBOLEV, S. L., On a theorem in functional analysis, *Mat. Sb. (N. S.)* **4** (**46**) (1938), 471–497 (Russian). English transl.: *Amer. Math. Soc. Translations* (2) **34** (1963), 39–68.

98. SOBOLEV, S. L., *Some Applications of Functional Analysis in Mathematical Physics*, Amer. Math. Soc., Providence, R. I., Third edition, 1991.

99. STEIN, E. M., *Singular Integrals and Differentiability Properties of Functions*, Princeton Univ. Press, Princeton, N. J., 1970.

100. STREDULINSKY, E., *Weighted Inequalities and Degenerate Elliptic Partial Differential Equations, Lecture Notes in Math.* **1074**, Springer-Verlag, Berlin–Heidelberg, 1984.

101. STRÖMBERG, J.-O. and WHEEDEN, R. L., Fractional integrals on weighted H^p and L^p spaces, *Trans. Amer. Math. Soc.* **287** (1985), 293–321.

102. TORCHINSKY, A., *Real-variable Methods in Harmonic Analysis*, Academic Press, San Diego, Calif., 1986.

103. TURESSON, B. O., *Nonlinear Potential Theory and weighted Sobolev Spaces*, Ph. D. Thesis no. 387, Dept. of Math., Linköping Univ., Linköping, 1995.

104. VODOP'YANOV, S. K., Weighted L_p potential theory on homogeneous groups, *Sibirsk. Mat. Zh.* **33**:2 (1992), 29–48 (Russian). English transl.: *Siberian Math. J.* **33** (1992), 201–218.

105. VODOP'YANOV, S. K., Thin sets in weighted potential theory, and degenerate elliptic equations, *Sibirsk. Mat. Zh.* **36**:1 (1995), 28–36 (Russian). English transl.: *Siberian Math. J.* **36** (1995), 24–32.

106. WALLIN, H., Continuous functions and potential theory, *Ark. Mat.* **5** (1963), 55–84.

107. WIENER, N., The Dirichlet problem, *J. Math. and Phys.* **3** (1924), 127–146.

108. ZIEMER, W. P., *Weakly Differentiable Functions*, Springer-Verlag, New York, 1989.

Index

Lecture Notes in Mathematics

For information about Vols. 1–1545
please contact your bookseller or Springer-Verlag

Vol. 1691: R. Bezrukavnikov, M. Finkelberg, V. Schecht-man, Factorizable Sheaves and Quantum Groups. X, 282 pages. 1998.

Vol. 1692: T. M. W. Eyre, Quantum Stochastic Calculus and Representations of Lie Superalgebras. IX, 138 pages. 1998.

Vol. 1694: A. Braides, Approximation of Free-Discontinuity Problems. XI, 149 pages. 1998.

Vol. 1695: D. J. Hartfiel, Markov Set-Chains. VIII, 131 pages. 1998.

Vol. 1696: E. Bouscaren (Ed.): Model Theory and Algebraic Geometry. XV, 211 pages. 1998.

Vol. 1697: B. Cockburn, C. Johnson, C.-W. Shu, E. Tadmor, Advanced Numerical Approximation of Nonlinear Hyperbolic Equations. Cetraro, Italy, 1997. Editor: A. Quarteroni. VII, 390 pages. 1998.

Vol. 1698: M. Bhattacharjee, D. Macpherson, R. G. Möller, P. Neumann, Notes on Infinite Permutation Groups. XI, 202 pages. 1998.

Vol. 1699: A. Inoue, Tomita-Takesaki Theory in Algebras of Unbounded Operators. VIII, 241 pages. 1998.

Vol. 1700: W. A. Woyczyński, Burgers-KPZ Turbulence, XI, 318 pages. 1998.

Vol. 1701: Ti-Jun Xiao, J. Liang, The Cauchy Problem of Higher Order Abstract Differential Equations, XII, 302 pages. 1998.

Vol. 1702: J. Ma, J. Yong, Forward-Backward Stochastic Differential Equations and Their Applications. XIII, 270 pages. 1999.

Vol. 1703: R. M. Dudley, R. Norvaiša, Differentiability of Six Operators on Nonsmooth Functions and p-Variation. VIII, 272 pages. 1999.

Vol. 1704: H. Tamanoi, Elliptic Genera and Vertex Opera-tor Super-Algebras. VI, 390 pages. 1999.

Vol. 1705: I. Nikolaev, E. Zhuzhoma, Flows in 2-dimensio-nal Manifolds. XIX, 294 pages. 1999.

Vol. 1706: S. Yu. Pilyugin, Shadowing in Dynamical Systems. XVII, 271 pages. 1999.

Vol. 1707: R. Pytlak, Numerical Methods for Optimal Control Problems with State Constraints. XV, 215 pages. 1999.

Vol. 1708: K. Zuo, Representations of Fundamental Groups of Algebraic Varieties. VII, 139 pages. 1999.

Vol. 1709: J. Azéma, M. Émery, M. Ledoux, M. Yor (Eds), Séminaire de Probabilités XXXIII. VIII, 418 pages. 1999.

Vol. 1710: M. Koecher, The Minnesota Notes on Jordan Algebras and Their Applications. IX, 173 pages. 1999.

Vol. 1711: W. Ricker, Operator Algebras Generated by Commuting Projections: A Vector Measure Approach. XVII, 159 pages. 1999.

Vol. 1712: N. Schwartz, J. J. Madden, Semi-algebraic Function Rings and Reflectors of Partially Ordered Rings. XI, 279 pages. 1999.

Vol. 1713: F. Bethuel, G. Huisken, S. Müller, K. Steffen, Calculus of Variations and Geometric Evolution Problems. Cetraro, 1996. Editors: S. Hildebrandt, M. Struwe. VII, 293 pages. 1999.

Vol. 1714: O. Diekmann, R. Durrett, K. P. Hadeler, P. K. Maini, H. L. Smith, Mathematics Inspired by Biology. Martina Franca, 1997. Editors: V. Capasso, O. Diekmann. VII, 268 pages. 1999.

Vol. 1715: N. V. Krylov, M. Röckner, J. Zabczyk, Stochastic PDE's and Kolmogorov Equations in Infinite Dimensions. Cetraro, 1998. Editor: G. Da Prato. VIII, 239 pages. 1999.

Vol. 1716: J. Coates, R. Greenberg, K. A. Ribet, K. Rubin, Arithmetic Theory of Elliptic Curves. Cetraro, 1997. Editor: C. Viola. VIII, 260 pages. 1999.

Vol. 1717: J. Bertoin, F. Martinelli, Y. Peres, Lectures on Probability Theory and Statistics. Saint-Flour, 1997. Edi-tor: P. Bernard. IX, 291 pages. 1999.

Vol. 1718: A. Eberle, Uniqueness and Non-Uniqueness of Semigroups Generated by Singular Diffusion Operators. VIII, 262 pages. 1999.

Vol. 1719: K. R. Meyer, Periodic Solutions of the N-Body Problem. IX, 144 pages. 1999.

Vol. 1720: D. Elworthy, Y. Le Jan, X-M. Li, On the Geo-metry of Diffusion Operators and Stochastic Flows. IV, 118 pages. 1999.

Vol. 1721: A. Iarrobino, V. Kanev, Power Sums, Gorenstein Algebras, and Determinantal Loci. XXVII, 345 pages. 1999.

Vol. 1722: R. McCutcheon, Elemental Methods in Ergodic Ramsey Theory. VI, 160 pages. 1999.

Vol. 1723: J. P. Croisille, C. Lebeau, Diffraction by an Immersed Elastic Wedge. VI, 134 pages. 1999.

Vol. 1724: V. N. Kolokoltsov, Semiclassical Analysis for Diffusions and Stochastic Processes. VIII, 347 pages. 2000.

Vol. 1725: D. A. Wolf-Gladrow, Lattice-Gas Cellular Automata and Lattice Boltzmann Models. IX, 308 pages. 2000.

Vol. 1726: V. Marić, Regular Variation and Differential Equations. X, 127 pages. 2000.

Vol. 1727: P. Kravanja, M. Van Barel, Computing the Zeros of Analytic Functions. VII, 111 pages. 2000.

Vol. 1728: K. Gatermann, Computer Algebra Methods for Equivariant Dynamical Systems. XV, 153 pages. 2000.

Vol. 1729: J. Azéma, M. Émery, M. Ledoux, M. Yor, Séminaire de Probabilités XXXIV. VI, 431 pages. 2000.

Vol. 1730: S. Graf, H. Luschgy, Foundations of Quantization for Probability Distributions. X, 230 pages. 2000.

Vol. 1731: T. Hsu, Quilts: Central Extensions, Braid Actions, and Finite Groups,. XII, 185 pages. 2000.

Vol. 1732: K. Keller, Invariant Factors, Julia Equivalences and the (Abstract) Mandelbrot Set. X, 206 pages. 2000.

Vol. 1733: K. Ritter, Average-Case Analysis of Numerical Problems. IX, 254 pages. 2000.

Vol. 1736: B. O. Turesson, Nonlinear Potential Theory and Weighted Sobolev Spaces. XIV, 173 pages. 2000.

4. Lecture Notes are printed by photo-offset from the master-copy delivered in camera-ready form by the authors. Springer-Verlag provides technical instructions for the preparation of manuscripts. Macro packages in T_EX, L^AT_EX2e, $L^AT_EX2.09$ are available from Springer's web-pages at

http://www.springer.de/math/authors/b-tex.html.

Careful preparation of the manuscripts will help keep production time short and ensure satisfactory appearance of the finished book.

The actual production of a Lecture Notes volume takes approximately 12 weeks.

5. Authors receive a total of 50 free copies of their volume, but no royalties. They are entitled to a discount of 33.3 % on the price of Springer books purchase for their personal use, if ordering directly from Springer-Verlag.

Commitment to publish is made by letter of intent rather than by signing a formal contract. Springer-Verlag secures the copyright for each volume. Authors are free to reuse material contained in their LNM volumes in later publications: A brief written (or e-mail) request for formal permission is sufficient.

Addresses:

Professor F. Takens, Mathematisch Instituut,
Rijksuniversiteit Groningen, Postbus 800,
9700 AV Groningen, The Netherlands
E-mail: F.Takens@math.rug.nl

Professor B. Teissier
Université Paris 7
UFR de Mathématiques
Equipe Géométrie et Dynamique
Case 7012
2 place Jussieu
75251 Paris Cedex 05
E-mail: Teissier@ens.fr

Springer-Verlag, Mathematics Editorial, Tiergartenstr. 17,
D-69121 Heidelberg, Germany,
Tel.: *49 (6221) 487-701
Fax: *49 (6221) 487-355
E-mail: lnm@Springer.de